Hazel Ros.... was bor.... various places across southern England .... She had a diverse education, attending nine dit.. —— and then St Hugh's College, Oxford, where she gradua.... chemistry.

She is married to Francis Rossotti, who was a member of the same research group at Oxford. Between 1954 and 1961 they worked together at the Royal Institute of Technology, Stockholm, and at Edinburgh University. They have also written a monograph together on chemical equilibria. They have a son and a daughter.

Since 1962, Dr Rossotti has been a chemistry tutor and a Fellow of St Anne's College, Oxford, and has also taught science part-time to senior boys at Summerfields Preparatory School, Oxford. In addition to research articles, she has written two text-books on branches of solution chemistry, three science books for children and *Introducing Chemistry* (Pelican Books, 1975). She is a keen photographer, mainly in black and white.

HAZEL ROSSOTTI
COLOUR

Princeton University Press
Princeton, New Jersey

Published by Princeton University Press,
41 William Street, Princeton, New Jersey 08540

Copyright © 1983 by Hazel Rossotti
All rights reserved
First Pelican original edition, 1983
First Princeton Paperback printing, with corrections, 1985

LCC 84 – 11451
ISBN 0 – 691 – 08369 – X
ISBN 0 – 691 – 02386 –7 (pbk.)

Reprinted by arrangement with Penguin Books Ltd.
Made and printed in Great Britain
by Richard Clay (The Chaucer Press) Ltd,
Bungay, Suffolk
Set in VIP Times

Clothbound editions of Princeton University Press books are printed
on acid-free paper, and binding materials are chosen for strength and durability.
Paperbacks, while satisfactory for personal collections, are not usually suitable
for library rebinding.

CONTENTS

Part Six: Uses and Links

LIST OF TEXT FIGURES

FOREWORD

The catalogue of the names of the many people who have contributed to this book looks like one of those lists of benefactors to a good cause. So even if Penguin would agree to its inclusion, it would make dreary reading, and I should be anxious lest I had not made it complete. For I am deeply grateful for all the help I have received from many quarters: from my husband and close family, from colleagues and other friends in a diversity of disciplines, or in none; from total strangers; and from former strangers who are now friends. I have picked the brains of some, but many more have volunteered suggestions of their own accord. Did I know about such and such an article, picture, word, exhibition, lecture, poem, book, phenomenon? Usually, I didn't, and much enjoyed following their leads, which took me down many unlikely pathways, to learn about fluorescent scorpions, textile designs, colour words in biblical Hebrew, new forms of street-lighting, the rainbow in Celtic mythology, Newton's influence on English poetry, the tattooing of seamen, wavelength discrimination in bees, Minoan wall-paintings and much else besides. When librarians produced books, and secretaries produced typescript, they usually produced ideas as well. I have learnt a great deal from them, as I have from the Master, staff and students of the Ruskin School of Drawing and Fine Art (in Oxford) and from all at the Museum of Colour (in Bradford). And, in particular, I learnt that almost everyone has some interest in some aspect of colour.

A few of this host of benefactors shall not remain nameless. Gerry Leach and Peter Wright encouraged the project from the start. They, Dimitris Gounelas, and the late Ted Bowen, FRS, were all good enough to read the whole text and to make detailed criticisms of it. Their many insightful comments, together with the wealth of suggestions from the unnamed multitude, have greatly improved the book. After all the care they took, its current blemishes must surely have crept in later. Only one suggestion did I totally ignore: Ted Bowen's laconic comment, 'p. 203, for "elephant" read "Dutchman"'; but, in our family sub-culture, it was always 'elephant'.

It is also a great pleasure to acknowledge all that Sydney Harry has taught me about colour, and his most generous gifts of pieces of his own work, including the jacket design for this book. I am grateful, too, to all those who made publication possible: to the editorial staff of Penguin Books, to their designers and their printers; and to all the copyright holders who have kindly allowed us to reproduce material. My debt to previous authors is obvious, immense, and gratefully acknowledged.

I am glad to have this chance to thank all these people very warmly for their help because I am indeed deeply grateful. I hope that the book, to which they contributed so much, succeeds in transmitting some of the enjoyment which their suggestions gave to me, and so gives to its readers something of the two great benefits I myself have obtained: those of interests shared and of visual awareness increased.

Hazel Rossotti
St Anne's College,
Oxford

PREFACE TO THE PRINCETON EDITION

I should like to thank John Mollon for telling me about the recent experimental results which form the basis of the revised Figures 47, 48 and 49, and of Tables 2 and 3. I am grateful too to the publishers for allowing me to incorporate some of the many good suggestions made by Alan Cowey, to whom I am extremely grateful for his generosity with both his time and his expertise on colour vision.

Hazel Rossotti
Oxford, July 1984

INTRODUCTION

Colour General definition.

A property of material objects, including sources of light, by which they are visually distinguished as possessing the qualities of redness, greenness, brownness, whiteness, greyness, etc.

– *Colour*, Report on Colour Terminology,
British Colour Group (1948), sect. IV:
'Terms Used in Ordinary Speech'

$$\text{Color} = (X, Y, Z)$$
$$\text{and } (X, Y, Z) = f_1(E, R, \bar{x}, \bar{y}, \bar{z})$$

Color, in colorimetry, is a concept which means exactly what is represented by (X, Y, Z), and (X, Y, Z) is computed as a functional relation of energy (E), reflectance or transmittance (R), and the amounts $(\bar{x}, \bar{y}, \bar{z})$ of three arbitrary lights required to match each part in turn of an equal-energy spectrum for a standard observer ... Depending on the viewing conditions, the surrounding objects or areas, the sizes and relative positions of objects, the adaptive state of the viewer, and a host of other things, a particular object that the colorimetrist characterizes by a given (X, Y, Z) can take on many different appearances. Some of these differences can be analyzed as differences in hue (redness, yellowness, greenness, blueness), saturation (dissimilarity of a given hue from a neutral of the same brightness), and brightness (the perceptual similarity of a hue-saturation combination to some one of a series of neutrals ranging from dark to light or dim to bright). The hues, saturations, and brightnesses that are abstracted from complete visual experiences and used to represent dimensions along which color may vary are functionally related to many things, among which are the spectral characteristics of the stimulating energy (E, R_λ) the spectral matching functions of a particular observer (r, g, b), the observer's memory (M) for similar objects, the surround (S), adaptive state of the observer (A), neighbouring objects (O), the observer's attitude at the moment (T), and so on (u, v, w).

$$\text{Color} = (H, S, B) = f_1(E, R, M, S, A, O, T, \bar{r}, \bar{g}, \bar{b}, \ldots, u, v, w)$$

A report of the Inter-Society Color Council
Subcommittee for Problem 20 (1963):
'Basic Elements of Color Education'

That which I am writing about so tediously, may be obvious to someone whose mind is less decrepit.

– WITTGENSTEIN, *Remarks on Colour* (295)

> Do not all charms fly
> At the mere touch of cold philosophy?
> There was an awful rainbow once in heaven:
> We know her woof, her texture: she is given
> In the dull catalogue of common things.
> Philosophy will clip an Angel's wings,
> Conquer all mysteries by rule and line,
> Empty the haunted air, and gnomèd mine –
> Unweave a rainbow.

– KEATS, *Lamia*, II, 229

'People,' wrote Goethe, 'experience a great delight in colour, generally.' 'People,' claims *The Oxford Companion to Art*, 'do not ordinarily make colours the object of special consideration.'

Both generalizations seem fair statements. Taken together, colour emerges as an enriching aspect of human experience, neither painstakingly analysed nor totally taken for granted; just enjoyed. So why the need for any books about it?

Of previous books on colour, some are treatises on such limited topics as the measurement of colour, colour photography, pigments and 'colour theory'. Others, of wider appeal, are directed towards those with particular interest in the scientific, artistic or architectural aspects of colour. This book has been written for a still larger group of readers, the lookers; for those who actively observe, rather than passively absorb, the colours around them.

Many such people, in their childhood, must have wondered about such problems as: Why do pebbles look brighter when wet? Is there a 'right' order in which to arrange a set of coloured crayons? Are blue rooms really 'cold'? Why do coloured rings appear at the edges of inkspots on a blotter? Where is the indigo in a rainbow? Why do some clothes change colour when you iron them? What happens to the colours on a top when you spin it? Why does the world look different through a red sweet-paper? What are the mauve streaks in bonfire flames? Why does iodine make black stains on a starched apron? What are the colours you see when you press your eyes? Why must you call a white horse 'grey'? What colours are soap-suds, dew-drops, black beetles, wood pigeons and the sky? But for the adult the basic colour problem is, of course: What is colour? When confronted with such a question, one turns for help to *The Oxford English Dictionary*, which states:

The particular colour of a body depends upon the molecular constitution of its surfaces, as determining the character and number of light vibrations it reflects.

This seems satisfactorily lucid, if somewhat prosaic, until one reads on:

Subjectively, colour may be viewed as the particular sensation produced by the stimulation of the optic nerve by particular light vibrations ...

And, at a still further remove, from any coloured object:

This sensation can be produced by other means, such as pressure on the eye-back or an electric current.

How subjective should we be? Is not red paint indisputably red? *The Oxford English Dictionary* quotes Reid, who, in 1764, stated succinctly:

Philosophers affirm that colour is not in bodies but in the mind; and the vulgar affirm that colour is not in the mind, but a quality of bodies.

Further progress through *The Oxford English Dictionary* brings us to 1856 and Ruskin's unanalysable assertion that 'colour is the most sacred element in all visible things'.

In the earlier part of the book, we shall, like Reid's 'the vulgar', take the common-sense view that colour is 'a quality of bodies', in particular that quality by which a body can affect the light which falls on it. So Part One of the book gives a brief outline of the nature of light, of matter and of some of the ways in which one may modify the other. Since, from the common-sense viewpoint, the colours which we see depend on the composition of the light which we receive from the object, Part Two describes the available sources of light as well as the wide variety of ways in which colour may be generated. Part Three gives, still from the objective, common-sense standpoint, a wide survey of the origin of colour in the natural world, mineral, vegetable and animal.

But we must recognize, with Reid's 'philosophers', that colour is a *sensation*, produced in the brain, by the light which enters the eye; and that while a sensation of a particular colour is usually triggered off by our eye receiving light of a particular composition, many other physiological and psychological factors also contribute. Part Four explores the way in which the eye responds to the composition of the light, and the effect of the many 'additional' factors on the sensation of colour. The difficult problem of devising a system to record sensations of colour is also discussed.

The rest of the book is devoted to different aspects of man's use of colour. Part Five deals with the technology, both of the reproduction of colour on screen and page, and of the dyes and pigments used to colour various surfaces and materials. The final section explores the diverse contexts in which colour is used to convey information and modify feelings. It surveys the relationship of colour to the visual arts, to music and to words.

It is hoped that, if this book succeeds in increasing awareness and understanding, then it will also enhance enjoyment. Some readers may, however, share Keats's view (page 14) that beauty can only be savoured in an atmosphere of ignorance. Those who prefer mystery to enlightenment may take heart. How fully do we yet understand the oscillations of a lightwave, or the forces within a drop of water, let alone the workings of the human eye and brain? We may have unwoven the rainbow, but how intimately do we know the strands?

PART ONE:
LIGHT AND DARK

And the light shineth in darkness; and the darkness comprehended it not.

– John 1:5

O say! What is this thing called Light?

– COLLEY CIBBER

We all *know* what light is; but it is not easy to *tell* what it is.

– DR JOHNSON

Give ear now to arguments that I have searched out with an effort that was also a delight. Do not imagine that white objects derive the snowy aspect they present to your eyes from white atoms, or that black objects are composed of a black element. And in general do not believe that anything owes the colour it displays to the fact that its atoms are tinted correspondingly. The primary particles of matter have no colour whatsoever, neither the same colour as the objects they compose nor a different one....

– LUCRETIUS

And the Colours generated by the ... two Prisms, will be mingled at PT, and there compound white. For if either Prism be taken away, the Colours made by the other will appear in that Place PT, and when the Prism is restored to its Place again, so that its Colours may there fall upon the Colours of the other, the Mixture of them both will restore the Whiteness.

– NEWTON

1

LIGHT PARTICLES

Church windows by day have little to offer the outside world. But, after dark, the glory of man's homage to his god shines outward, leaving the congregation with windows no more colourful than the face of a switched-off digital watch. Why this dramatic contrast between light falling on to stained glass and light streaming through it? What happens to the light when it meets a window? Or to the window when light passes through it? How much do we actually know about light?

One thing we do know about light is that, without it, there is no colour. 'All colours,' wrote Bacon in his essay on religion, 'will agree in the dark.' It is not clear what sort of darkness Bacon had in mind. He may, for literary emphasis, merely have been stating the obvious: in total darkness, we can see nothing at all, neither shape nor colour. But we are very seldom in absolute darkness. Usually the dark is light enough for us to distinguish areas of different shades of grey which we can piece together to give us some idea of shapes, even though the scene is far too murky for us to guess at colours. We normally see colours only if the light is fairly bright. But where does light come from?

The light which streams into a cathedral in daylight owes its origin to the changes which take place in the sun, where small particles are combining to form larger, more stable ones, and an enormous amount of energy is being set free. Some of this energy, which floods our solar system, is of a form which we perceive through our eyes. We call it light. The light which shines out of the cathedral during winter evensong has less dramatic an origin, though it, too, is given out only when some substance has more energy than it can retain. This extra energy was previously always obtained by combustion: of candles, torches, oil or gas. If the flame contained specks of soot, the heat set free made them glow and give out light; or the heat lit up an incandescent gas mantle. Nowadays energy is usually pumped into a lamp as electricity, to heat a wire filament or initiate changes in a fluorescent tube. But the materials cannot accommodate this extra energy and so dispose of it exceedingly rapidly in the form of light.

And what is this light? Unless, like Cibber's beggar boy, we are blind, we all know what light is. In our present technological age, it can surely no longer be true that none can *tell* what it is? Scientific knowledge has taken so many large paces forward since Dr Johnson pronounced light, like poetry, to be recognizable but indescribable. Although light could, of course, be seen, its workings could not then be visualized. But this was not for want of trying. Scientific argument about it has flared and subsided since Newton considered light to be a stream of particles. Others, like Huygens, believed it to be a succession of waves. Which of these was right?

For the non-physicist, the nature of light is still very far from clear. Perhaps the nearest we can get to understanding it is to envisage a series of rhythmical electrical disturbances in space. These disturbances pulsate much like a wave; so it looks as though Huygens was right. But, unlike sound waves or water waves, light 'waves' need no medium in which to oscillate. And, in other respects, light behaves like a stream of particles. So maybe Newton was right. But the particles (called photons) are very different from those in a burst of grapeshot, a jet of water or a puff of wind. For one thing, they owe their mass, and indeed their very existence, to their movement. They travel through empty space at about 300,000 km per second: but if they could be brought to a standstill, they would simply cease to exist. Another peculiarity of the photon is its unchartability. We cannot say exactly where it is at any given moment, although we can assess the chance that it is in a certain place at a particular time.

So it might seem even more difficult to tell what light is than it seemed in Dr Johnson's day. A beam of light cannot be fully described either as a wave or as a stream of particles, though it has some features of both. In a way, both Huygens and Newton were right.

For most of our discussion of colour, we can think of light as a stream of minute, elusive particles which create a pulsating electrical disturbance. The chances of finding one of these photons at a certain distance from the source of light varies with time, in much the same way as does the movement of the crest of a water wave at a particular place in a pond after the waters have been disturbed.

What happens when these photons, whether from the sun or from a light bulb, meet a stained-glass window? Something must surely happen because, without the photons, we see nothing. When the window is lit, do the colours we see come from the light, or from the window; or from the interaction of the two? Why does the window seem so drab when light *falls on* it, but glow with splendour when light *shines through* it? We clearly need to know more about those elusive 'motes' which make up a beam of light. They cannot, surely, all be the same. Candlelight looks very different from sunlight.

A single photon can differ from another photon in only one respect: its energy. We might have guessed that a high-energy photon would travel faster than a lower energy one. But this is not so. In empty space, all photons travel at exactly the same speed. However, all photons are slowed down when they travel through air, or water, or glass; and those of different energy are slowed down to different extents.

Although the energy of a photon does not affect its speed of travel through a vacuum, it does affect the frequency of the oscillatory disturbance. We might also guess that, the more energetic the photon, the greater the number of oscillations performed in a given time; and now we would be right. We can illustrate this by the wavy lines in Figure 1, where the crests

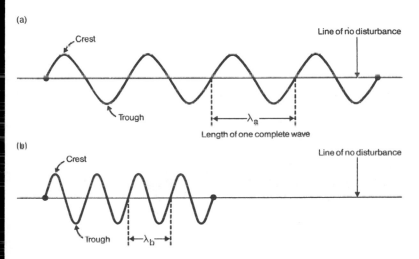

Figure 1. Light waves *Representation of light as a wave of electromagnetic disturbance. The wavelength of wave (a) is twice that of wave (b). For each wave, four complete cycles are shown.*

represent the highest chance of finding a particular type of electrical disturbance while the troughs represent the same chance of finding a second type of disturbance which opposes the first sort. Between each trough and crest the wave crosses the line representing zero, and at these points there is no disturbance. In Figure 1, the distance between two consecutive crests in the wave (a) is twice that in the more energetic wave (b); so wave (b) goes through two complete cycles (crest-zero-trough-zero-crest-zero-trough-zero-crest) in the same distance in which (a) goes through only one. We can use the distance between two successive crests as a label which specifies the leisureliness of oscillation. It is called wavelength and often written as lambda (λ). The units are nanometres, written as nm, where one nanometre is one thousandth millionth part of a metre. A high value of this lambda

denotes considerable sluggishness and so implies a low frequency of oscillation and a low energy. A photon of shorter wavelength oscillates more
frequently and carries more energy. And to these differences in energy our
eyes respond, enabling us to *see colours*. High-energy light, in which most of
the photons have wavelengths of around 400 nm, looks blue or violet, while
low-energy light, containing photons mainly of wavelengths around
700 nm, looks red. We can easily produce a stream of photons, nearly all of
which have the same wavelength (of 589 nm), by dropping a pinch of salt
over a flame: and we see the same yellowish-orange glow in many of our
street lights. The main colours which are produced by photons of different
(single) wavelengths are shown in Table 1.

But the world we normally see is not illuminated by strongly coloured
lights. In daylight, and in indoor artificial light, white paper looks white;
under a sodium street lamp, however, we might well say it looked yellow. In
daylight, of course, we are viewing the paper by 'white' light. As we shall
see, we can only perceive true whiteness if both the light and the paper can
themselves be described as 'white'.

Table 1. *Visible Light Waves*

Colour sensation	Wavelength (nm)
Red	760 to 647
Orange	647 to 585
Yellow	585 to 575
Green	575 to 491
Blue	491 to 424
*Violet	424 to 380

* Red light performs about 440 million million oscillations every second, while violet light
performs about 730 million million in the same time.

What then is the 'white' light, emitted by the sun? White does not feature
in the colours produced by photons of any particular wavelength; these all
produce single-colour, or so-called 'monochromatic', light. The white light
given out by the sun, and the near-colourless light produced by many
artificial sources, contains a more motley collection of photons, which carry
a wide variety of energies. Sunlight contains photons of all energies in the
visible range (of wavelengths 380 to 760 nm) as well as those with energies
both above and below those to which our eyes respond.

With a prism, of either glass or water, we can show that white light is a
mixture: we can split sunlight into its rainbow components (see Figures 2
and 3). We do not need to duplicate Newton's darkened room, through

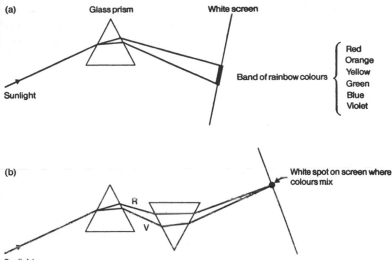

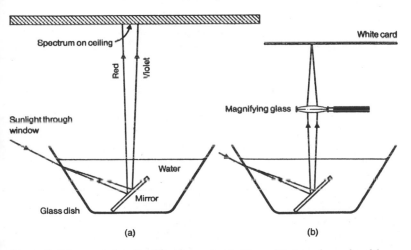

Figure 2. Prismatic colours *A glass prism can split sunlight into light of different wavelengths, which can be reflected as a coloured band from a white screen (a) or recombined into white light with a second prism (b).*

Figure 3. A home-made prism *The glass prism in Figure 2 (a) can be replaced by a water prism and a mirror (a). A magnifying glass (b) can be used instead of the second prism in Figure 2(b) to focus light of different wavelengths on the same place, and so to give a spot of white light.*

which a thin stream of sunlight entered through a small circular hole he had cut in the shutters. But Newton did a second, even more important experiment with the 'coloured ghost' (or 'spectrum') into which his prism had transformed the sunlight. With a second prism, he recombined the rainbow colours he had made. And from this mixture he obtained: white light.

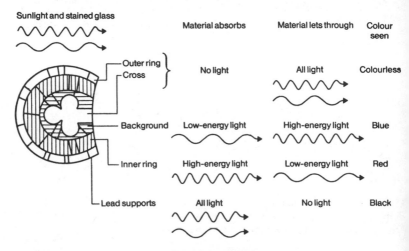

Figure 4. From a rose window *The effect of stained glass on sunlight, which for clarity is represented as a simple mixture of high-energy light and low-energy light. The roundel is one of twelve surrounding Christ in Majesty in the south rose window at Chartres.*

So what happens when this 'white' sunlight, this light of diverse wavelengths, passes through a stained-glass window – a rose window, maybe, of areas of deep blue and deep red, made all the richer by thin strips of clear glass between them (see Figure 4)? The blue glass absorbs many of the photons which fall on it; but not all. And it is choosey about which, having a very marked preference for those of low energy (and so of long wavelength). The result is that it mops up most of those which would give us a sensation of green, and almost all of those which we would perceive as yellow, orange and red. But it allows free passage to most of those of higher energy; and so light of the lower wavelengths passes through and we see a deep, rich blue. The ruby glass is equally selective, but unlike the blue glass, it absorbs photons of high and medium energy, letting through only light of the highest visible wavelengths; those which we see as deep red. The colourless glass is transparent to nearly all the light which falls on it, regardless of its wavelength. Sunlight streams through it almost unchanged.

So, while clear glass allows free access to all photons, stained glass of a particular colour acts as a filter, as a type of security barrier, allowing passage to some of the photons, but not to others, according to their energy.

How is this magnificent selectivity achieved? And why can we enjoy it only from its more dimly lit side? To understand what takes place when sunlight falls on glass, we need to know not only about the sunlight. We must also try to envisage what is happening inside the glass, and at its surface. Let us start with a pane of clear, colourless, window glass.

2

WHITE LIGHT ON CLEAR GLASS

Large plate-glass doors are often adorned with opaque heraldic motives, or divided horizontally by a bar. These are no mere decorations: they serve to alert us to the presence of the glass. Without them, people commonly bump into, or even crash through, the almost invisible glass. Light passes through a perfect piece of glass almost as easily as it does through air itself. And it is only those objects which do affect light in some way that we can see.

But a pane of glass will be almost invisible only if it is faultless: the sides must be absolutely flat, and perfectly parallel, and there must be no hint of any ridge or knot. Imperfections will become apparent with the slightest movement of the eye; an object seen through the glass will look rippled, like a reflection in a stream. In an old window, it is easy to guess that those panes which give us a clear unbroken view are modern replacements; any old glass is likely to be of uneven thickness and so to produce a willowy image.

Seen from the outside, the glass of an unlit window looks blackish; but shiny. We may see the curtains at either side, but the central part is very dark. In it we may see a sombre reflection of the world outside, partly obliterated by a much lighter area, the highlight. But colourless glass is not invariably shiny. Pictures may be framed under glass which, although perfectly transparent, is dully non-reflective. Rougher matt surfaces are produced by glass engravers, often in designs of great complexity. And when colourless window-pane glass is ground to powder, or spun into glass wool, it no longer seems colourless and transparent, but rather a gleaming, opaque white.

How is it that ordinary colourless, almost invisible glass can present us with so many different faces? And what clues can they give us about how light behaves when it meets glass? Since we can see clearly through a window pane, photons must be able to travel through it. If the sheet of glass is faultless, the light appears to travel in a straight line between the object and the observer; thus we can see through it as clearly as if it were not there. Uneven panes, or glass which has been moulded into ribbed or frosted patterns, give distorted, interrupted or totally fragmented images. This occurs because a ray of light is bent whenever it passes from air into glass, except when it enters exactly perpendicularly to the surface. (The same thing happens when light travels from air into water, which is why a straight stick, partly dipped into water, appears bent at the surface.) When the light emerges from glass into air, it is bent again, in the opposite direction. If the glass is a perfect sheet, the two bends exactly counteract each other, so the glass does not cause any change in the direction of the light, which appears to have travelled in a straight line from the object (although it has, in fact, been very slightly shifted to one side, as Figure 5 shows). However, if

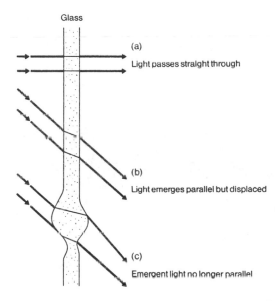

Figure 5. How light travels through glass *The passage of light through clear, colourless glass. The rays are initially parallel. (a) Light falls perpendicularly on perfect sheet glass. (b) Light strikes perfect sheet glass at an angle. (c) Light falls on a fault in the glass.*

the surface at which the light leaves the glass is not parallel to the one at which it entered, the two bends do not cancel each other out, and the light does not travel in a straight line between the object and the observer. We see the world inverted and minute through a glass marble; and through a reeded glass door as a piece of abstract weaving. Distortion has been much exploited by makers of telescopes, microscopes, spectacles, cameras and window roundels.

But not all of the light which falls on glass is able to cross the surface;

some bounces back, and as with a ball on a tennis court, it will leave the surface at the same angle as that at which it struck it. So a smooth surface will reflect light regularly, and if it is illuminated mainly from one direction, we will see a reflection of the source as a highlight, the shininess emphasizing the smoothness of the glass. If the light which falls on the window does not come directly from a light source, but has been bounced off a near-by building or tree, we see reflections of these objects in the glass; sombre, but well-defined.

We do not, however, see clear reflections, or shiny highlights, on the rough surfaces of etched, or of non-reflecting, glass. The light bounces off these surfaces, too, at the same angle at which it meets them, but in these glasses each part of the surface is set at an angle to its neighbour. So, as the light bounces off the glass, it is scattered in a variety of directions (see Figure 6); and the surface looks dull.

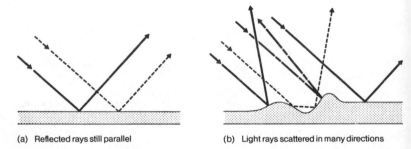

(a) Reflected rays still parallel   (b) Light rays scattered in many directions

Figure 6. How light bounces off glass *The reflection of light from (a) a smooth surface and (b) a rough surface. The rays are initially parallel.*

It is easy to 'see why windows look so dark from the outside, and why reflections in them are more sombre than those in polished metal or metal-backed mirrors. We see only that light which enters our eye, and so, from outside a house, we see only that which bounced back towards us off the glass, whether by regular reflection or random scattering. Most of the light which falls on glass does not, of course, bounce back. Window panes are transparent, and most of the light which falls on them travels onwards, into the house, away from the observer. The window frames are often themselves bright, maybe of satinized aluminium or painted in white gloss. Both types of surface reflect or scatter nearly all the light which falls on them, making the windows seem even darker by contrast (see Figure 7).

How is it that most of the photons which fall on glass travel through it,

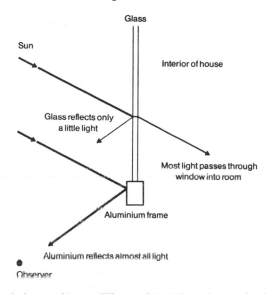

Figure 7. The dark eyes of houses *When sunlight falls on the outside of a window, the glass usually looks darker than the window frame.*

while a few seem to be rejected? What happens inside the glass when photons strike it? What is the interior of glass like?

A three-dimensional chart of the inside of a piece of glass, or indeed of all other sorts of matter, would show three different regions: tiny, very dense atomic nuclei, which carry a positive electric charge and account for almost all the weight and the material; regions containing clouds of electrons, which are much lighter particles, each carrying a negative electric charge; and empty space. Electrons are much heavier than photons, but resemble them in elusiveness. We cannot say exactly where one will be at any instant. So, in our chart, the boundaries between the electron clouds and the empty space cannot be exactly defined.

Everything around us is built from these very dense nuclei, each embedded in a cloud of electrons, much like a piece of lead shot in the centre of a vast sphere of candy floss. There are over one hundred types of nuclei, each with an electron cloud to 'fit' it, the criteria of a good fit being a number of electrons which gives the entity a very low, or zero, electric charge and allows, if possible, the outer 'wrapping' of electrons to be completed. Although there are constraints about how many neighbours, and of what sort, each of these entities is likely to have, a very large number of stable arrangements is viable. The entities may exist singly (as 'atoms'), in combinations of varying sizes, ranging from pairs (containing either identical or dissimilar nuclei) and small groups and clusters of various shapes to larger chains, rings, branched structures and infinite three-dimensional fibres, networks and arrays. Some resemble piles of cannon balls, others vast climbing frames or twisted rope. In glass itself, about two thirds of the nuclei are those of oxygen. The rest are mainly those of silicon, but if soda has been added to soften the glass, or borax to harden it, some other nuclei, such as sodium or boron, will also be present. Glass has an untidy structure, not unlike old chicken wire or chain-link fence, as if many of the connections have rusted away, and the others have been misshapen by compressing the sheet or netting into a very rough ball.

How are the components of matter held together? What are the remaining 'links' in the chicken-wire structure of glass?

A collection of atomic nuclei and electrons settles down into the most stable arrangement possible; and the most stable arrangement, like that of even more complex systems, such as human societies, is determined by various opposing tensions. Because each of the nuclei carries the same type of charge, one repels another; but electrons and nuclei, being of opposite charge, attract each other. Moreover, an arrangement is more likely to survive if the outer layer of electrons is complete. So it is not surprising that the components most affected by the presence of their neighbours are the outermost electrons around each nucleus; and it is these which, in effect, hold the assembly together.

In an isolated atom of any chemical element, the positive charge on the nucleus is exactly balanced by the negative charge on the electron cloud; but the outer layer of this cloud may be incomplete, and, if so, the atom will not be very stable. If it encounters other suitable atoms, it is likely to combine with them, to give some aggregate in which the outer electron clouds are reshuffled and which is more stable than the separate components. We can think of the electron cloud around a nucleus as being made up of a number of completed shells (represented by pink candyfloss) surrounded by the outermost (white candyfloss) layer, which may or may not be complete (see Figure 8). In most of the materials around us, these outer clouds are no longer neatly apportioned to particular nuclei, but are common to two or more of them. In glass and diamond, they are concentrated between the components, like clasped hands (see Figure 9). But in metals (see Figure 10), in which the components are even more closely packed together, the outer electron clouds have seeped into all the spaces between the components, much as treacle might flow into all the interstices within a heap of ball bearings. It is because of the difference between 'clasp' and 'treacle' electron clouds that glass and diamond look so very unlike aluminium or gold.

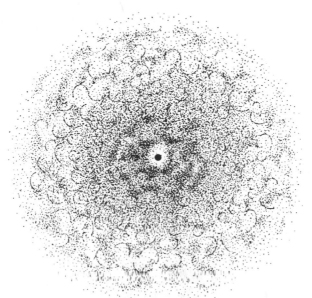

Figure 8. A single atom of matter *Lead-shot and candyfloss model.*

It is not surprising that light, which is an electrical disturbance, should interact with a cloud of electrically charged particles; nor that its effect depends both on its energy and on the nature of the electron cloud.

Suppose that a stream of photons of energy in the visual range meets a sheet of clear glass. The oscillating electrical disturbance of the light makes the electron clouds oscillate; they give out energy of the same wavelength as the original light, in all directions. And as each electron cloud has a number of close neighbours, they pulsate in unison. So each electron cloud acts as a tiny source of light. What happens when the light given out by one cloud meets that emitted from another? When the trough of one wave exactly coincides with the crest of another, two effects cancel each other out, as Figure 11a shows. We can also see, from Figure 11b, that when the crests of two waves exactly coincide, they reinforce each other; and the light doubles in intensity. This is what happens in the direction in which the light is travelling. But, in all other directions, the oscillations of neighbouring electron clouds counteract each other, and the two waves are obliterated. Hence, in daylight, light travels unimpeded through a window pane into a room, provided that the glass is flat, smooth and clean. An observer inside

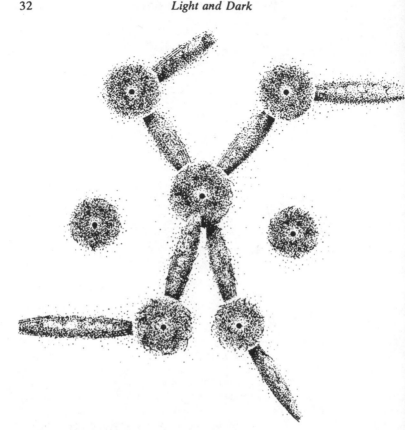

Figure 9. Glass *Impressionistic sketch of its open 'clasp-type' structure.*

the room therefore has a clear view of the world outside; but to an observer outside, the window seems unfathomably dark.

Although a beam of light which strikes the glass perpendicularly travels forward in exactly the same direction, it does so at a slightly reduced speed. But when the light slants on to the glass, there is a change both in speed and in direction; the beam is bent as it crosses the surface. We can envisage that edge of the beam which meets the glass first as being slowed down the sooner. The exact angle of such a refraction depends on the wavelength of the light and on the two substances through which it is travelling. Light is bent more on passing from air to diamond, and less on passing from water to glass, than it is on crossing the boundary between air and glass.

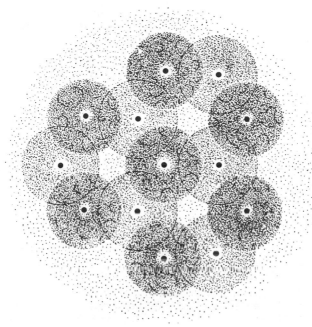

Figure 10. A metal *Impressionistic sketch of the close-packed 'treacle-type' structure of a metal.*

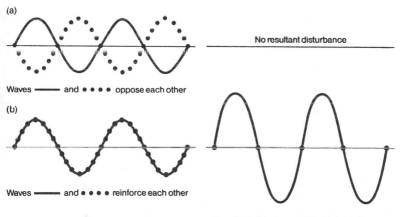

Resultant disturbance of double intensity

Figure 11. Interference of light *Two photons of the same energy interact, and may (a) obliterate or (b) reinforce each other.*

As we saw earlier (page 28), not all the light which arrives at the glass is able to cross the surface. Some is reflected. When the electron clouds, stimulated by a stream of photons, oscillate in unison, any light emitted backwards or sideways will be cancelled out by the oscillations of neighbouring clouds which obstruct them. But an electron cloud on the surface of the glass has no neighbour behind it, and so there is no obstruction to prevent back scattering. Since a smooth pane of glass has only a very small proportion of its electron clouds on the surface, only a small proportion of those photons which strike it are re-emitted towards the observer. Reflections, though clear, are rather weak. If a pane of glass is ground to a coarse powder, a much higher proportion of the material is now unobstructed, on some edges, orientated higgledy-piggledy with respect to a large number of edges of adjacent grains of glass. A much higher percentage of the light is scattered, but in random directions. The powder looks a dull white, but we see no clear reflections. We can easily increase the proportion of 'edge' electron clouds, and hence the proportion of light scattered, by grinding the glass to a finer powder. The smaller the particles, the whiter it looks. A gloss-painted white window frame and a matt, whitewashed wall also owe their appearance to the scattering of light by fine particles, as we shall see in Chapter 8.

Aluminium window frames look quite different from white gloss-painted ones, even though both reflect much of the light which falls on them. But then metals have quite different outer electron clouds from those in paint or powdered glass. More precisely, within a piece of metal, there is one single outer electron cloud, extending over the whole sample. This 'treacle-type' continuous cloud is much more sensitive to electrical disturbance than the more localized 'clasp' types of cloud in glass. When exposed to light, it oscillates so violently that none of the photons can penetrate further into the body of the metal. All the light is therefore scattered backwards: metals are totally opaque, but brightly shiny; and so, if smooth, give bright and well-defined reflections.

So far we have seen how white light can interact with clear glass, with white pigments and with a white metal. Even within this limited repertoire, remarkable variety can be achieved. The light can be transmitted straight, or transmitted bent. It can be reflected regularly, or scattered randomly; so that a surface may look shiny, or matt. The same pane of glass can look almost black, or be invisible; but if it is crushed, it becomes a white powder. But, as yet, we have not needed to invoke our knowledge that white light is a mixture of photons of different energies: of lights of different colours.

In the next chapter, we shall see how this heterogeneity of white light can give rise to colour.

PART TWO:

LIGHTS AND COLOURS

When this comes in contact with other objects, it may pass through, as it does in particular through glass.

– L U C R E T I U S

Which plainly shows, that the Lights of several Colours are more and more refrangible one than another, in this Order of their Colours, red, orange, yellow, green, blue, indigo, deep violet; ...

– N E W T O N

Hitherto I have produced Whiteness by mixing the Colours of Prisms. If now the Colours of natural Bodies are to be mingled, let Water a little thicken'd with Soap be agitated to raise a Froth, and after that Froth has stood a little, there will appear to one that shall view it intently various Colours every where in the Surfaces of the several Bubbles; but to one that shall go so far off, that he cannot distinguish the Colours from one another, the whole Froth will grow white with a perfect Whiteness.

– N E W T O N

By a change of light, according as the beams strike it vertically or aslant, ... a peacock's tail, profusely illumined, changes colour as it is turned this way or that. These colours, then, are created by a particular incidence of light. Hence, no light, no colour.

– L U C R E T I U S

... do not all fix'd Bodies, when heated beyond a certain degree, emit Light and shine?

– N E W T O N

> 'Yes,' I answered her last night;
> 'No,' this morning, sir, I say,
> Colours seen by candle-light
> Will not look the same by day.

– E L I Z A B E T H B A R R E T T B R O W N I N G

Now sight is primarily the perception of colour, but along with the colour, it discriminates that body which has colour ... also whether it is rough or smooth ... whether its composition is ... wet or dry.

– J O H N O F D A M A S C U S
Of the Orthodox Faith

3

STEADY COLOURS

A red, white and blue flag flutters on the church flagstaff. Raised at dawn, and to be lowered at dusk, it is flown only in daylight. If any passer-by were asked whether the flag seemed to change colour, he would probably answer with an amazed, 'No, of course not.' The colours seem to be the same red, white and blue whether the day is bright or dull, windy or still; and from whatever spot our observer is standing. But as we shall see in later chapters, the question is not so silly as it sounds (nor the answer so simple).

If the passer by were to go into the church, he would find the interior dark compared with the daylight outside. The windows, though large, are mainly of stained glass; had the glass been colourless, the church would have seemed much lighter. But the coloured windows glow with richness, even when the sun is not shining directly through them. One old window is mainly of deeply coloured glass, royal blue and ruby, separated by small colourless areas. Another window, slightly more recent, also includes panes of green, yellow and purple.

The colours of stained-glass windows are as unchanging as those of the flag (or even more so, since they persist unfaded for centuries). When the sunlight streams through them, they do, of course, glow even more vividly than when the sky is overcast; but the blue still looks royal, as the red still looks ruby, when the observer changes his position, or when a cloud passes over the sun. And, so far as most people's colour memory can recall, the colours still look the same when seen from outside a lighted church by night. We are so used to this apparent unchangeability of many of the colours around us that we are tempted to regard it as obvious; even self-evident. We say that something 'is', or 'has', a certain colour; but we shall see that this is frequently an oversimplification.

How is it, though, that an object can 'have' a colour? The very darkness of churches with stained-glass windows confirms that the coloured glass absorbs much of the sunlight which falls on it. We have seen (page 24) that ruby glass absorbs much of the high-energy, low-wavelength light, from the blue violet end of the spectrum through the turquoise, green and yellow regions, right up to the orange. Only the lowest-energy visible light, the long wavelength deep red, passes freely through the glass. The blue glass, on the other hand, absorbs much of the low-energy red, orange and yellow light, but lets through most of the blue light together with some violet and some green light from neighbouring regions of energy. Green glass absorbs light at both ends of the visible spectrum, but is transparent to light of intermediate energy; to green light. No wonder that stained-glass windows make a church seem rather dark: they let through only a fraction of that light which can pass through a colourless pane of the same size.

Stained-glass windows can contain other colours, too, such as yellow and purple. We shall discuss in Chapter 15 the regions of the spectrum from which glass of these colours absorbs light.

The colours in the flag are produced in exactly the same way as are those in a stained glass window. The different parts of the flag absorb light of different wavelengths from the heterogeneous white light which falls on them. But while the glass is transparent to the light which it does not absorb, the flag is opaque. Glass transmits the non-absorbed light, while cloth scatters it back towards the observer. So the three regions of a red, white and blue flag scatter long wavelengths, all wavelengths and short wavelengths. Green cloth would scatter light of mainly intermediate wavelength; and black would absorb all the light which falls on it, so that very little returns to the observer.

Of all those things around us which 'have' a particular colour, the great majority owe their colour to their ability to absorb light of some energies more readily than light of others, while their opacity depends on what happens to such light as is not absorbed. Photographic filters and colour slides, clear toothbrush handles and transparent cellulose wrapping, allow, like stained-glass windows, the unabsorbed light to travel on unhindered. Light of selected wavelengths is also transmitted through the coloured covers of traffic signals and car indicator lights, but it is bent at various angles as it crosses the ridged surface of the lamp covers: the light which emerges is somewhat diffused and we cannot see a clear image of the bulb.

Opaque materials scatter rather than transmit any light which they do not absorb; and they, too, differ in the extent to which they diffuse the light which reaches the observer. Highly polished gold scatters in a very regular way and gives clean, bright reflections. Worn, finely scratched gold scatters more randomly, giving diffuse and ill-defined reflections. Most opaque materials scatter still more haphazardly. Their highlights, if any, are unimpressive, and may go unnoticed by those who have not been trained to see them. And if light were scattered totally at random, the surface would appear completely matt, like blotting paper or a dry 'undercoat' of paint.

We know how a clear material such as plain glass can transmit and reflect light; how a finely powdered one scatters almost all of it; and how a polished metal reflects almost all of it. But how does stained glass absorb light of some wavelengths but transmit that of others; or coloured cloth absorb some colours from the spectrum and scatter the rest? Why do so many things 'have' a particular colour?

Photons, as we know, are carriers of energy; indeed, they *are* a form of energy. And we have seen that, when photons come into contact with some material, their energy is often 'lent', very briefly, to the material: the energy of the photons is absorbed by the electron clouds, which then oscillate, re-emitting photons, most of which are transmitted forward through the glass window, or scattered backwards from powdered glass or polished metal.

In coloured materials, the energy of some of the photons is transferred to the electron clouds in a different way. Normally, the electron clouds have the lowest possible energy, and this is true both for the electrons closely associated with one particular nucleus (the 'pink candyfloss' in our model on page 30) and the outermost ('white candyfloss') electron clouds which are often associated with two or more of the components. Sometimes, however, the outer clouds, or the outermost electrons in the inner regions, may absorb energy and rearrange themselves into a less stable, though momentarily viable, structure. They soon lose this extra energy and revert to their previous, most stable arrangement. But they do not usually* dispose of it by re-emitting photons; the excess energy seeps away through the material as heat, causing a very slight increase in temperature. Two questions spring to mind. If electron clouds can absorb light without re-emitting it, why do they absorb only those photons of particular energy rather than absorbing photons of all energies? (If they absorbed light of all colours, the material would, of course, look black.) And since some materials can absorb light of certain energies, why cannot all materials do so?

A given material can absorb photons only of particular energies because each arrangement of an electron cloud contains a specified amount of energy. Thus the change from the most stable (lowest-energy) state to a viable alternative can be accomplished only if the cloud increases its energy by exactly the right amount.† We might envisage an electron as a gymnast about to leap from a secure footing on the floor on to a narrow bar. He jumps with precisely controlled energy; a leap with more momentum, or with less, would be unsuccessful. We may extend the analogy slightly. The bar is moving backwards and forwards from the gymnast, over a short distance. At one moment, he will need to use a slightly different amount of energy from that needed an instant previously. Moreover, some jumps may

* See, however, page 58.

† The exactness is such that the international standard of length is now defined in terms of the wavelength of light required for one such change and the standard of time in terms of the frequency of another. The *metre* is the length: equal to 1 650 763·73 wavelengths in vacuum of the radiation corresponding to the transition between the levels $2p_{10}$ and $5d_5$ of the krypton-86 atom. The *second* is the duration of 9 192 631 770 periods of the radiation corresponding to the transition between the two hyperfine levels of the ground state of the caesium-133 atom.

be more difficult to achieve than others, and so be undertaken less frequently. The difficulty of the feat does not depend on the energy needed. A near-by bar 60° above the gymnast might seem a less inviting target than a much more distant one at 15°.

Electrons often absorb photons within a narrow band of wavelengths, because the energy needed to make a particular jump depends on the positions of neighbouring clouds and nuclei, which alter as these vibrate. And some electronic changes are, indeed, more likely to occur than others: the more probable is the change, the greater is the absorption, and so the purer is the remaining light and the stronger the colour we see.

Why is it that glass, diamond, salt, polythene, chalk, cellulose, chromium and ice all seem to lack colour? Are they incapable of absorbing energy? Perhaps there are no other viable alternatives for their electron clouds? No. There are, in fact, possible alternative arrangements of electron clouds, and some of these substances do absorb energy from sunlight. But the energy gap between the most stable arrangement and the lowest-energy alternative is quite large. High-energy photons are needed to bring about the change. So those which are absorbed from sunlight are those of very low wavelength; too low for us to see. Clear glass, for example, absorbs ultra-violet light from the sun, but almost no visible light.

Substances that absorb coloured light are those whose electron clouds can undergo changes which need rather small amounts of energy. Strong colours are usually the result of changes in an electron cloud which is shared between a number of components. Many dyestuffs, both natural and synthetic, contain large groups of components with a large communal outer electron cloud which can be somewhat reshuffled by absorbing small amounts of energy. Hence those substances subtract some visible light from sunlight. The changes are of high probability and are therefore intense. Many of the colours found in flower petals and leaves are of this type, as are many of the synthetic dyestuffs made from aniline and other derivatives of coal tar. The bright turquoise blue (Figure 12a) characteristic of many older Pelican bindings and the crimson of our own red blood corpuscles (Figure 12b) are both members of a family of coloured substances in which the frameworks are somewhat like four-petalled flowers, sandwiched between two continuous electron clouds. At the centre of these molecules is a metal atom; the blue dye contains copper, and our haemoglobin contains iron.

In some of the stronger earth colours, such as rust and ochre, part of the outer electron cloud can be shifted from the periphery of one nucleus to that of another. This, too, needs only the amount of energy which visible light can supply. Silver tarnishes much more quickly in industrial areas and in houses with solid fuel than in regions where the air is less contaminated with

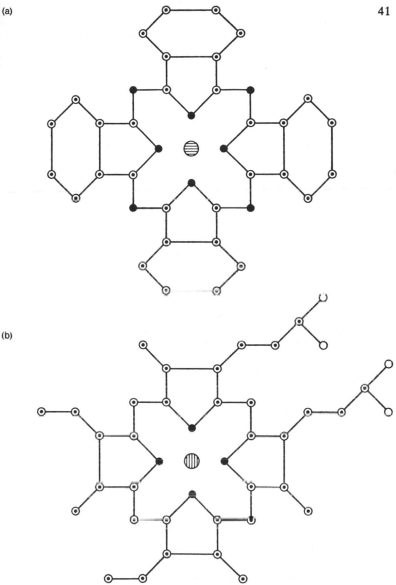

Figure 12. Printer's blue and blood red *Framework of structures of (a) the turquoise dyestuff copper phthallocyanine and (b) haemin, obtained from red blood corpuscles, showing positions of the nuclei of carbon ⊙, nitrogen ●, oxygen ○, copper ⊜ and iron ⦿. Most of the outer electron cloud lies either along the lines, or above and below them.*

sulphur fumes. The colour again arises from a partial shift of the outer electron cloud; this time from sulphur to silver.

The very intense colour of Prussian blue is produced by a transfer of charge over a considerable distance, from corner to corner of a 'climbing-frame' structure, in which the edge of each cube is a cyanide group (the pair of atoms C-N with one extra electron) and the corners are iron atoms which have lost, alternately, either two or three electrons. Only a little energy is required to transfer electrons from one iron atom to another via an inter-mediate cyanide group. The intense blue-black of iodine stains on starch are thought to arise from easy transfer of electrons along chains of iodine atoms, trapped inside the long tunnels which are formed, like the insides of spiral springs, from the starch.

A rather different change is responsible for many of the colours of stained glass, and of pottery glazes. When visible light is absorbed, some readjust-ment of electrons takes place at the outer edge of one of the inner (unshared) electron clouds. The colours of many gem-stones (Chapter 8), of green copper roofs and of blue copper sulphate also arise in this way.

So the characteristic colour of an object, in daylight, depends on the wavelength needed to produce a readjustment of electrons, because it is these energies which determine the composition of that remaining mixture of light which enters our eye and causes the sensation of colour.

We have said that many objects and materials 'have' a particular colour. If we postpone our discussion of the complex relationship between the light which we receive and the sensation we experience, how far is it true that the same material will always relay to us the same mixture of wavelengths? The light which reaches us depends, of course, on the composition of that light which falls on the object. Here we are assuming that our source of light is daylight (or artificial lighting of very similar composition). We shall discuss other sources of light in Chapter 6.

If we keep the lighting the same, surely the colours of objects are unchangeable? Usually this is so; but not always. Some dyed cloth changes colour alarmingly under a hot iron; the possible arrangements of electrons in the dye vary with the temperature, and therefore so does the energy needed for rearrangement. Luckily, the change is rapidly reversible. The red spot on the handle of an electrically heated hairbrush works in the same way. When the brush is hot and ready for use, the spot is nearly black.

Similarly, reversible changes in colour can be caused by light itself. Light-sensitive sunglasses and aircraft cockpits are made from transparent, colourless materials which darken reversibly on exposure to bright light. As with temperature-sensitive substances, the extra energy absorbed increases the number of ways in which yet more energy can be taken up. A greater amount of visible light is then absorbed, and the colour darkens. When the intensity of the light falls, the sunglasses or cockpits revert to their initial paler form. But the most important reversible light-sensitive colour change is surely one which we cannot see: it takes place within our own eyes and is the basis of the sense of sight (see Chapter 11).

Can we say that the colour of a sheet of coloured cellophane changes when it is folded double? It is not surprising that two layers of coloured transparent material, like cellophane or glass, look a deeper colour than a single layer. A block of coloured glass looks the deepest colour if viewed through the greatest thickness, because the light then encounters the greatest number of absorbing electron clouds before it reaches the eye. In the same way, the colour of whisky gets progressively paler as water is added to it, although there is no change in the quality of the colour.

However, we get quite a different result if we add water to the juice of blackcurrants, bilberries or red cabbage. The *quality* of the colour changes; from purplish-red to blue. As we add water, we naturally dilute all the particles in the juice: those which affect the acidity as well as those which absorb visible light. And it is this change in the acidity of the juice which alters the energy of the electron clouds in these vegetable dyes and hence the colour of the liquid (see Chapter 9).

If we break a piece of stained glass, do we change its colour? Suppose we put a pane of red glass on a piece of white cardboard and then tread on it. Maybe we merely break it into a hundred pieces. The colour would, of course, be unchanged. But if we now stamp on it, and break it into smithereens, it will look no longer deep red, but pink. There is a great increase in surface area, several angled sides being provided by each chip of glass, so that much of the light which falls on the glass is scattered, regardless of its wavelength. A much lower proportion can pass through the glass, be reflected off the white cardboard and pass through the glass again, emerging as predominantly long-wavelength 'red' light. The smaller the chips, the higher is the ratio of scattered white light to transmitted red light, and the paler the pink colour of the crushed glass. If we pulverize it with a hammer, it becomes almost white, though we have destroyed none of the red colouring matter.

However, despite the sensitivity of some electron clouds to temperature and to acidity, and despite the increased proportion of white light scattered as a coloured solid is crushed, the colours which arise by absorption are moderately constant. A piece of red glass looks red if you hold it up to the light, if you hold it up to a white wall and if you move it between the two.

But there are other ways of coaxing colour from sunlight. As we shall see in the next chapter, some of them give colours which are far from constant: shimmering colours which seem to change with the merest movement of the head.

4

SHIMMERING COLOURS

A spider's web, slung with dew, on a winter morning. Colourless drops on a colourless thread; until the sun comes out. Then both web and dew light up with brilliant colours which change with the gentlest breeze, with the slightest movement of your head. Why should such threads and drops, colourless in diffuse daylight, become coloured when the light is more strongly directional? Water in a tumbler does not usually look coloured, so why should the water of a dew-drop do so? Do we see shimmering colours elsewhere? And are the colours which shine out from the dew-drop the same as those we see from the web?

These elusive, changing colours are by no means unusual; but they differ among themselves. Some, like those from the dew-drop, comprise all the colours of the rainbow. We can see the same sequence, of red, orange, yellow, green, turquoise, blue, violet in the bevelled edges of mirrors, in the sparkling of a chandelier or of a diamond, or in the silvery tape with which many young people adorned their vehicles in the late 1970s. Spiders' web colours are quite different. We see them, too, in bubbles, on gramophone records, and from sunlight on a frosty windscreen. But we see them best from oily patches on wet roads, an intense iridescence, uncertain, it seems, as to whether to shine out peacock blue, or purple, or resplendent gold. And what of a real peacock? As we shall see in Chapter 10, shimmering colours, often within a more restricted range, abound in the animal kingdom, from peacock to starling, from abalone to mussel, and from butterfly to beetle.

How do these colours arise? Why do they keep changing? And why does a spherical dew-drop look so different from a spherical bubble?

The dew-drop is the simplest. It acts exactly like the triangular glass prism with which Newton and his predecessors split sunlight into its components, each of a different colour. If we stand in one position, we may be at exactly the right angle for the green light to reach our eye. A slight movement one way and we will see first yellow, then orange and red: and a movement in the opposite direction will allow us to see, first, the greenish-blue light, then the deeper violet blue. How can a dew-drop, or a prism, disperse the colours in this way? We know that light is bent or 'refracted' when it crosses a boundary between one transparent medium to another, because it is slowed down when it enters a denser medium. The angle at which light is bent on passing from air to water or to glass varies slightly with wavelengths because, although all light travels at the same speed in a vacuum, light of high energy is slowed down slightly more when it crosses into a denser material. So when white light passes from air to water, or to glass, the blue is bent the most markedly, and the red the least. We saw in Chapter 2 that, when white light travels through a pane of glass (or thicker block with parallel faces), the angle at which the light is bent on emerging from the block exactly counteracts the bend it suffered on entering. And as this is true for light of any wavelength, the light which emerges from a perfectly transparent block has the same composition as that which entered it; sunlight is not split into colours. However, since neither a triangular prism, nor a spherical drop, has parallel sides, the light emerges from them at an angle different from that at which it entered. Light is bent on passing through the denser material, and light of different colours is bent at different angles. Sunlight is splayed out into a continuous spectrum of colour. Movement of the head allows the eye to meet the light of different wavelengths emerging at different angles, and so to catch the various colours in turn.

The strands of a spider's web are much narrower than a dew-drop. Indeed, the purple, gold and peacock colours are often produced by very fine layers of material: a thin shell of soap solution enclosing a bubble of air, or a film of oil on water. This family of shimmering colours is produced by transparent layers which are only as thick as a few wavelengths of visible light. Imagine a beam of light of one particular wavelength, shining on to a bubble or on to water which is covered by an exceedingly thin layer of oil. Most of the light will, of course, pass through the film to the air inside the bubble or to the water beneath the oil, depending on how much each material bends light. But some will be reflected back from the first boundary, and some more from the second one. An observer, in the right position, receives light resulting from both reflections. The two streams of photons will interact with one another. The path of the one which has been (twice) through the oil is, of course, longer than that which has not. And if the difference between the two path lengths happens to be an exact number of half wavelengths, the crest of one wave will coincide with either the crest or the trough of the other. When a crest and a trough coincide, the two waves will cancel each other out (see page 31 and Figure 13). The difference in the distance through which the two reflected rays travel depends, of course, on the thickness of the liquid film; but it also depends on the angle through which the light travels from source to film to eye, and this in turn varies as the observer moves his head. In one position, the difference in the light paths will be just right for the red light beams to interfere with each other and be cancelled out; at another angle, green light will be removed, and at yet another, the same will happen to blue, and so on. So we receive mixtures of light of composition white-minus-red, white-minus-green or white-minus-blue; and we see peacock, purple or gold, shimmering exotically into each other with the slightest movement or change in film thickness.

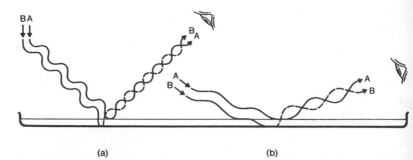

(a)   (b)

Figure 13. Oil patches and soap bubbles *Production of colours by interference of two light rays reflected from the upper (A) and lower (B) surfaces of a thin layer of liquid. As cancelling occurs when the distance travelled by the two rays differs by one wavelength, the high-energy light (a) is destroyed when the film is viewed from a high angle, while the low-energy light (b) disappears when it is seen from a lower angle.*

The gentler colours of soap bubbles and spiders' webs owe their delicacy to the fact that the 'subtractive' colours, of purple, gold and peacock, are much diluted with white light. Goethe observed that stronger colours can be seen in the bubbles of frothy chocolate than in soap suds: they have been diluted less. Although bubbles on a coffee filter are even more impressive, the most splendid interference colours of all are surely those of oil on wet tarmac, which reflects almost no white light. Gramophone records, viewed from a low angle, may show similarly iridescent colours, caused by the interference of light reflected from the top and bottom of the groove. The iridescence of ancient lead oxide glass is due to the interaction of light reflected from different layers which form as the glass deteriorates.

Very thin layers of tarnish on polished metal are often iridescent. Car radiators, spouts of metal coffee pots and reflectors of old electric fires may develop soap-bubble colours through years of contact between hot metal and the oxygen in the air. We may see similar colours when a penknife blade or needle is held in a flame. Although the oxidized layer is much thinner than the wavelengths of visible light, there is a change in 'phase' when light is reflected from a metal surface; a wave which meets the metal surface as a 'crest' leaves it as a trough. But since there is no such change when light meets a particle of oxide, interference still occurs and we see 'subtractive' colours. The variously coloured iridescence of an oxide layer formed on titanium has been exploited by present-day designers of jewellery and small pictures.

Not all shifting surface colours are caused by interference and the consequent loss of light of some wavelengths. Some are mere highlight effects, caused by variation in the amount of white light which is reflected. We saw in Chapter 1 how very sombre stained glass appears when seen only by the light reflected from it. When we look along a nap of velvet, the fibres are lying flatter than if we look against the nap; more light, of all wavelengths, is reflected from the smooth surfaces, giving pale, shining highlights. Against the nap, we see a more broken surface. Of the decreased amount of light reaching us, a higher proportion has been rejected by the dye; so the velvet looks duller and darker.

When a solid is so intensely coloured that it absorbs almost all the light of some wavelengths, the remaining light reflected from the surface at a glancing angle will have a bronzed appearance of white light minus that colour. If we view dried red ink from a low angle, it has a greenish lustre. As Goethe also observed, 'The bright red Spanish rouge exhibits on its surface a perfectly green, metallic shine,' and, 'All good indigo exhibits a copper-colour in its fracture.' For both fine cloth and dried ink, the proportion of light which is precisely reflected, rather than randomly scattered, varies with the angle of vision, giving these materials, too, a shimmering appearance.

Most of the more familiar 'kinetic colours', whether from a chandelier, an expanding bubble or rustling taffeta, are the result of one of the three effects we have mentioned here: the dispersion or fanning out of white light into its component colours, the subtraction of one colour from white light by interference in a thin layer, or the effect of changing highlights, whether white or coloured.

There are yet other ways in which colours can be coaxed from white light, only to vary richly with some small movement. But as these are less often encountered, and are slightly more complex in origin, they will be treated together, as 'special effects', in the chapter which follows.

5

SPECIAL EFFECTS

Look one night through a woven handkerchief at some distant street lamps, and you will probably see a number of separate dots. If the light is bright yellow, from sodium lamps, you will see a pattern of bright yellow dots; but if the light is whitish, you will probably find that the dots are striped, or slashed, with different colours. A near-by white lamp, seen through a piece of loosely woven cloth, appears to be surrounded with a series of circular haloes, often of 'oil-film' colours. The colours are produced from white light, but not from light from sodium lamps because white light is, of course, a mixture of a wide variety of wavelengths, whereas sodium lamps give out light over a very narrow band of wavelengths (see Chapter 7). And the colours are formed because the distance between two adjacent threads of the handkerchief is within, or not much greater than, the range of wavelengths of visible light. Although, for most practical purposes, light travels in straight lines, some of the photons in a beam are deflected when the light meets the edge of an opaque object, as happens when the light passes through a hole, or through a slit. It is as if the small gap through which the light passes were itself emitting photons, radiating light in all directions. Including straight on; for many of the photons proceed without deflection. But the rest diverge from this line, their paths being bent, outwards, at the edge of the hole or slit. These photons are 'diffracted'.

When light passes through two adjacent gaps which are separated by only a wavelength or so, we can guess that the photons emerging from one gap would interact with those from the other. At some positions they would reinforce each other, and at certain others be annihilated, depending on how the distance between the gaps is related to the wavelength of the light (see Figure 14). Thus, if we pass white light through a very closely spaced pair of slits, we should expect to see some oil-film colours, because, at some particular position, one colour would have been removed from the white light, and at a different position, some other colour would have been cancelled out. Instead of two slits, we may have large numbers of them, very close together (say, 10,000 per centimetre), like a very fine grating. There will then be extensive interference between neighbouring waves. At any one position, light of only one very narrow band of wavelengths, where the crests of the wave reinforce rather than annihilate each other, survives; at an adjacent point, the surviving photons are of a slightly different wavelength. So the white light is splayed out into rainbow colours. The grating produces a spectrum like that obtained from a prism. When you move either your head or the grating, a different wavelength reaches the eye and the perceived colour changes.

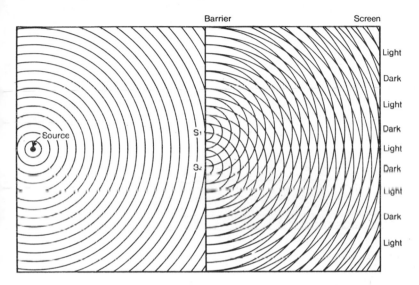

Figure 14. Diffraction by two slits *Light of a single wavelength reaches two slits S_1 and S_2 which each behave as sources. The resultant beams interfere, so that one crest may coincide with other crests to give bright regions, or with troughs to give dark ones. (Reproduced with permission from* Physics, Physical Science Study Committee, *D. C. Heath & Co., Boston, Mass., 1960.)*

'Gratings' are not always made up of alternate opaque and transparent lines, like minuscule straight iron railings. The transparent stripes can be replaced by other regularly spaced features which behave as if each were an independent source. Scratches, or alternating planes, can replace the black lines, and an opaque reflective surface, such as aluminium, can be used instead of a transparent material. Diffraction gratings can be reproduced cheaply and well, in plastic, from a base of aluminium, indented with angled grooves (see Figure 15). It is these which produce the vivid colours of some metal-backed pop-art trinkets and of the 'jewel tape' which adorns some cars and motor cycles; as the vehicle turns or drives past, we see a succession of rainbow colours.

Anyone who has two pieces of Polaroid and some Sellotape can make some very beautiful (and permanent) 'special effects' by sticking overlapping pieces of Sellotape on a transparent base. When the Sellotape design is placed between the two pieces of Polaroid and held up to the light, it has acquired a range of colours, each of which corresponds to a particular number of layers of Sellotape. If one of the pieces of Polaroid is then rotated, the colours change. A very impressive picture of a saint, built up from layers of Sellotape on glass, is exhibited in the Science Museum in London. Viewed through one sheet of Polaroid, it is almost invisible. But

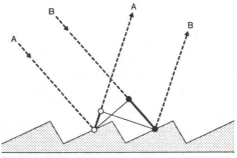

Grooves on flat aluminium

Figure 15. Diffraction by a grating *Reflection of light rays from a grooved aluminium. The distance travelled by ray A is less than that travelled by ray B by an amount equal to the difference between the two lengths* ●——● *and* ○——○. *A strong reflection is obtained when this difference is an exact multiple of the wavelength of the light.*

when a second sheet is added, the saint looks like a stained-glass window. The second Polaroid screen is then turned through a right angle, and the saint is again transformed, into a totally different stained-glass window. Any area which first let through only red light is now transparent to all colours but red, and so looks turquoise; and so on.

To understand how these colours appear, we must look at one more aspect of light waves: the extent to which a wave consisting of several photons is squashed flat. Does one crest of this composite wave exactly follow its predecessor, so that a series of crests travels in a straight line? Or does one crest follow another at an angle, so that a series resembles a cylindrical steel spring? Or is it more like a spring which has been only somewhat squashed? In fact, normal light, whether natural or artificial, is 'unpolarized' and follows no regular pattern. But it can sometimes be split into two 'plane-polarized' waves, at right angles to each other (see Figure 16). Here the crests follow linear paths. We can envisage one of the plane-polarized waves oscillating from left to right within the plane of this page, as if the crests are following one line of letters, the troughs another, lower one. The second plane-polarized wave would also oscillate from left to right, but both the crests and the troughs would follow the same line of type. The troughs would travel somewhat below the page, and the crests at the same distance above it.

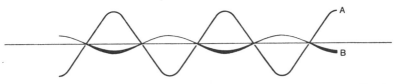

Figure 16. Polarization of light *Two plane-polarized waves (A) moving in the plane of the paper, and (B), at right angles to it, above and below the page.*

Since our eyes, unlike those of some insects, do not respond to differences in the plane in which light is polarized, things look the same in light which is polarized in one plane, polarized in another, or not polarized at all. But some materials, such as Sellotape, can change the plane of polarization on the light, and hence can produce colours when placed between sheets of Polaroid. Such substances often contain long particles, like rods or plates, in a regular arrangement. The structure of one face of a natural crystal, or of a machined block, will then differ from that of an adjacent surface. An extreme case might be a solid in which the rod-like molecules were arrayed like the teeth of a number of parallel combs (see Figure 17). The assembly would look quite different if viewed from the side, from the end or from above. When unpolarized light meets a directional material of this type, it will be split into two component waves which oscillate at right angles to each other

Polaroid itself has this type of structure, and it absorbs most of the visible spectrum, all except that light which is polarized in one particular plane, determined by its own orientation. It acts rather like the vertical bars of a lion's cage; a rib of beef on a fork can be poked into the cage only when it, too, is vertical. So, if the polarized light which emerges from a sheet of Polaroid is confronted with a second sheet of the same material, it will be transmitted only if the orientation of the second sheet is identical to that of the first; when the second sheet is at right angles to the first one, only a very little light (at the violet end of the spectrum) is transmitted. Such light as does get through a pair of 'crossed' Polaroid sheets therefore looks very dark purple, as can be checked by using two eyepieces from Polaroid sunglasses.

Interesting colour effects can be obtained when transparent slices of other materials are placed between two sheets of Polaroid. The first

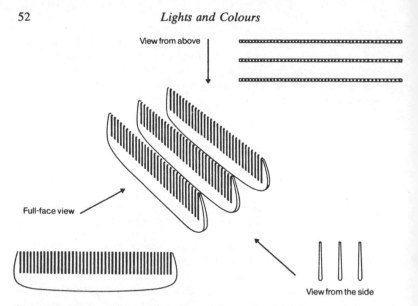

Figure 17. Three combs *An illustration of the way in which the gaps in an ordered structure seem to differ with its orientation.*

Polaroid produces plane-polarized light, which then travels through the sample. If the sample, too, contains rod-like or plate-like particles, or indeed has any features which vary with orientation, it will probably turn the plane in which the light is polarized. We can find out if it does so by seeing how the emergent light is affected by the orientation of the second sheet of Polaroid. When the wavelength of the light bears a particular relationship to the thickness of the sample, light of that wavelength may be cancelled out; and this, of course, produces 'oil-film' colours. Sellotape contains long, regularly orientated particles, and so is able to produce such colours, which vary with the number of layers of tape used.

These colours, produced when thin slices of material are placed between two polarizing plates, have been used to investigate the structures of many natural objects such as crystals, rocks and hailstones. Since small changes in structure give rise to different colours, the technique can be used to detect strain in plastics, glass and rubber. When sections of plastic models of bridges or of Gothic cathedrals are subjected to stress and viewed between polarizers, regions of different strain show up in different colours.

Orientation effects are also responsible both for those desk thermometers which indicate temperature by shining the appropriate number, each in a different colour, and, less colourfully, for the ubiquitous LCD output of calculators and digital watches. Both types of device contain substances known paradoxically as 'liquid crystals', which usually contain long particles. Like matches in a box, or logs floating down a river, these particles have a strong tendency to be aligned parallel to each other; but they can flow, either as an orientated group, or individually. Some liquid crystals formed by derivatives of cholesterol consist of layers of aligned 'logs', with each layer stacked slightly skew on the one below it, so that the whole assembly is twisted like a screw (see Figure 18). When unpolarized light falls on this type of liquid crystal, it is split into two beams, polarized at right angles to each other. One beam passes through the sample, while the

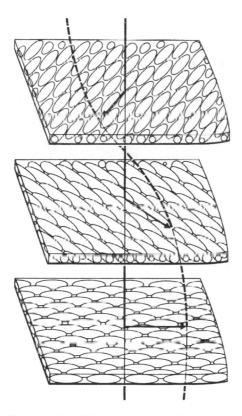

Figure 18. A highly ordered fluid *Three layers from a 'liquid crystal' of the cholesterol type, showing how the alignment of the 'logs' moves spirally around the vertical axis. (Reprinted with permission from G. H. Heilmeier,* Scientific American,. *vol. 222, No. 4, 1970, p. 100.)*

other is reflected from this surface. Light reflected from the exposed edges of different planes of aligned molecules may be cancelled out, producing iridescence akin to that of bubbles, but impressively intense. The colour is not affected by the thickness of the sample, but depends on the viewing angle and is particularly sensitive to the structure of the liquid crystal. The arrangement of particles is much more mobile than that in solid crystals and is disturbed by very small changes, which may give rise to dramatic changes in colour. Some liquid crystals change through the whole visible spectrum over a 1°C change in temperature. Since some liquid crystals recover quickly after a disturbance, and some slowly, they can be used in recording thermometers and other 'memory' devices. Stress also affects structure, and even slight pressure can cause a change in colour, which is exploited in such frivolous novelties as coasters, in which liquid crystals are covered by flexible transparent plastic; the coasters change colour when pressed, or even under the weight of a glass. Films of liquid crystals are used to detect imperfection in solid surfaces; since flaws alter the molecular arrangement, they show up a different colour. Liquid crystals may also be used for chemical analysis since molecules of any absorbed vapour and other impurities cause changes in the structure and hence in the colour. Coloured liquid crystals can also be used to translate invisible radiation into visible signals. Ultra-violet light may cause structural changes in substances of this type by breaking down some of the molecules, while infra-red radiation causes local increases in temperature. So colour 'maps' of both types of radiation are obtained.

The sombre, grey and black liquid crystal displays in calculators and digital watches contain liquid crystals in which the particles are again elongated and aligned; but they lack the skew orientation. Although colourless, they, too, are very sensitive to minute disturbances. A very small electric field can change in the extent to which light is bent on passing into the material and so can produce large changes in opacity. Liquid crystals, both coloured and colourless, have immense technological potential, mainly because so little energy is needed to promote minor changes in structure, which produce major changes in appearance. Their dramatic properties are not yet fully understood and are probably far from fully exploited. Yet more special effects will doubtless be available soon.

6

LIGHTS

Our main question, so far, has been, 'How do things look in daylight?' But we know how colours seem to change with the light. Clothes bought under artificial light may look subtly different when taken to a window and viewed by daylight; and sodium lights reduce a coloured scene to a near mono-chrome in yellow, khaki and black.

Such transformations are unsurprising. The sensation of a particular colour is produced by the exact composition of the light which enters our eye; and this naturally depends on the composition of the light which falls on the object, as well as on the way in which the object may alter this light. But how does the composition of light vary from one source to the next? And, even more fundamentally, how is light produced? Why are photons ever emitted? Whatever can neon signs, candles, light bulbs, glow-worms, television tubes, lightning and lasers have in common with the sun and stars, or with each other? Should we include luminous paint and fluorescent paper among our sources of light? (They certainly both shine more brightly than their surroundings.)

In all these cases, light is emitted by matter which has acquired more energy than it can contain. The difference in appearance of the light from the various sources depends on the rate of emission of photons in each part of the visible range. How, though, do these sources of light first acquire their excess energy?

Before the introduction of electricity, almost all sources of light were undergoing some irreversible change. By a rearrangement of particles, new substances are produced, and these normally contain less energy than did the substances from which they were formed. So energy is given out during this reshuffling process, sometimes in the form of photons, sometimes in other forms, such as heat.

The most dramatic changes are those which take place in the sun and stars, and result, through the fusion together of even smaller particles, in the formation of atomic nuclei (the very small, very dense particles, rep-resented by lead shot in our candyfloss model on page 30). The loss of energy is enormous, and accounts for the intensity of sunlight. We describe light as 'white' when its composition is the same as, or very similar to, that of light from our own sun. Light from other stars may be appreciably different in composition, depending on the reactions taking place within them. Although we are too far away to detect the differences in colour with the naked eye, we can see them in colour photographs as well as record them with instruments in the observatory.

The changes which give rise to candlelight are on a much smaller scale. The atomic nuclei remain intact, and indeed keep most of their enveloping electron clouds. But when vaporized candle grease combines with the oxygen in the air to give steam, carbon dioxide and soot, the nuclei and their immediate electron clouds become separated from their original neighbours and acquire new ones. The outermost electrons settle down into new, more stable arrangements and energy is evolved, almost entirely in the form of heat. The carbon particles in the soot become so hot that they glow, at first red and then, as the temperature becomes higher still, turning orange and then yellow; but they never acquire enough energy to emit green or blue light as well. So candlelight, unlike sunlight, is yellow rather than white. Light emitted by burning other carbon-rich materials, such as oil and coal, is a similar mixture of wavelengths.

Light can also be produced by substances which have been artificially supplied with energy. This extra energy may be pumped in as heat, as in a gas lamp, where the energy set free by the burning gas heats up the mantle, which then glows. So a gas lamp works on the same principle as a candle flame, except that the substance which glows in the lamp is a stationary artefact, while the soot in the candle flame is a mass of moving granules, themselves produced by the combustion.

Additional energy is, however, much more often supplied as electricity, which can be used to produce light in a number of different ways. In a traditional electric-light bulb, an alternating electric current passes through a fine tungsten wire and thereby heats it. The tungsten filament re-emits the energy over a wide range of visible and infra-red wavelengths; but the proportion of light from the blue, high-energy end of the spectrum is rather lower than in sunlight, and so tungsten lighting is yellower than daylight (see Figure 19).

Electricity can also be used to pump energy into matter without any gross increase in temperature. There are various types of tube containing gases at very low pressure, and if a high enough voltage is applied to such tubes, some electrons may be stripped off the gas particles. In certain lamps, the electrons then recombine with the emission of light, which may contain only a few, quite definite, wavelengths. Such lamps glow with colours which are characteristic of the gas used, as in red neon signs, and in deep yellow (sodium) street lamps. The red glow of a sodium lamp before it has 'struck' is due to the presence of neon. When the lamp is first switched on, it acts as a neon light, because at normal temperatures neon is a gas and sodium is a

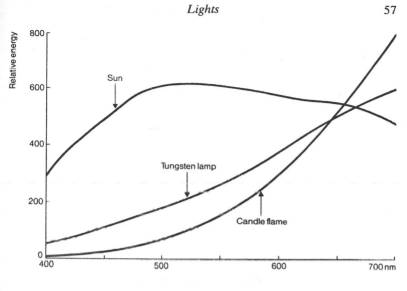

Figure 19. Light sources *Distribution of energies in some common light sources.* *(Adapted, with permission, from W. D. Wright,* The Measurement of Colour, *4th edition, Adam Hilger, London, 1969.)*

solid. But the heat generated by the neon lamp soon vaporizes the sodium, which then emits yellow light, which is much more intense than the red light from the neon.

Whiter street lighting is obtained from mercury lamps (which emit photons of a number of wavelengths, giving a pale bluish light), and from lamps which contain sodium vapour at such high pressure that the electron clouds of one gas particle are influenced by those of its neighbours. Since this allows for a much larger number of ways of disposing of energy, photons are emitted with a wide range of wavelengths and the light produced is of a soft, pale peach colour.

Electrons stripped from particles in gases may also be used to transfer energy to some light-emitting material inside the tube. The inner surfaces of fluorescent lighting tubes are coated with such 'phosphors' which emit light of a wide variety of wavelengths when struck by electrons; the tubes in black-and-white television sets work in the same way. Colour television tubes contain three different phosphors, each of which gives out light of a particular narrow range of wavelengths (corresponding to red, green and blue light) when activated by electrons (see Chapter 16).

Some solids can be made to glow by the passage of only a very small electric current through them; no ionized gas is needed as an intermediate and the solids do not become appreciably heated. These materials are 'semi-conductors' of electricity. Although, unlike metals, they do not normally have a continuous cloud of mobile electrons, only a little added energy is needed to promote an electron to form part of a mobile, conducting cloud: to remove an electron from a 'clasp'-type to a 'treacle'-type consortium. In some semi-conductors, light is given out when the electron falls back into its more stable 'clasp'-type state. These materials are widely used as Light Emitting Diodes (LEDs for short) for the displays of digital clocks and cash registers. The colour of a particular display depends on the energy gap between the 'clasp' state and the conducting 'treacle' level. Early LEDs were usually either green or red; blue ones were developed more recently.

Light itself can sometimes be used to provide a substance with excess energy and to stimulate from it the emission of further photons. Some substances or mixtures can absorb light in such a way that, in an appreciable number of particles, electrons acquire extra energy which they do not immediately lose. We might suppose that these particles would emit a photon of a particular energy and revert to a more stable state; but in practice the majority do not, unless they are stimulated to do so by collision with an oncoming photon of the same energy. The energy is temporarily stored. So if the substance is first excited by light or other energy, and then irradiated by a weak beam of light of frequency corresponding to energy difference between the excited and stable species, an intense pulse of photons of that same frequency is suddenly emitted. In practice, the necessary stimulus is provided by harvesting the photons emitted by the spontaneous decay of the minority of species which do not store their energy. The intensity of incoming Light has been Amplified by Stimulated Emission of Radiation, and the beam emitted by the resulting LASER consists of photons of effectively single energy (see Figure 20). Lasers operating at a large number of frequencies are now available and have numerous applications in medicine, technology, pure research and art.

What about such weak emitters of light as luminous paint and fluorescent dyes? If these are also 'sources' of light, how do they acquire their excess energy? Like lasers, they are excited, not by chemical or electrical energy, but by photons themselves. Unlike lasers, however, they re-emit photons of their own accord. Luminous paint is, indeed, a source of light. If a luminous clock has been recently illuminated, it will glow in the dark, with decreasing brightness, for several hours. It loses its excess energy by a very leisurely emission of photons. Fluorescent materials do not glow in total darkness, but they may appear to glow in the half-light. They often absorb, not only some visible light, but also photons of higher energy from the invisible,

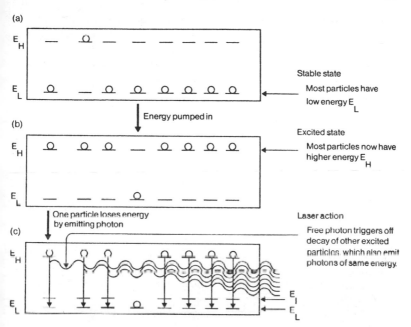

Figure 20. Laser light *(a) The original sample; (b) the storage of energy; (c) release of laser light. The energy of the light may be the difference (E_H-E_L) in energy between the stable and excited states of the photons; or the particles may emit photons of lower energy (E_H-E_1), and decay from the intermediate-energy state E_1 to the low-energy state E_L by emitting heat. (Adapted, with permission, from Atkins,* Physical Chemistry, *Oxford University Press, 2nd edition, 1982).*

ultra-violet region of the spectrum. When a fluorescent dye loses its excess energy, some of it is used to increase movement of the particles, and the rest is emitted very rapidly as light. So the photons emitted are of lower energy than those which were absorbed; and those from the ultra-violet range now come into the visible range. For some of the invisible photons absorbed, visible light is emitted. Thus the dye seems to give out more light than is falling on it; and it glows. Such dyes are much used for road-safety devices and in advertising. Some substances seem to be mainly one colour but to show a fluorescence of some other colour. For example, if a solution of dyestuff absorbs high-energy, violet light, it will look yellow (see page 179); and if some of the energy of the purple light makes the particles of the dye vibrate more violently, light of longer wavelength, e.g. green, will be given out. The yellow solution will then have a green fluorescent glow. Tonic water contains traces of quinine, which absorbs only in the ultra-violet but re-emits with a purplish fluorescence, which can just be seen in a gin and tonic placed in sunlight.

No wonder the quality of light depends so much on its source. On the one hand, there is the sun, emitting light over the whole range, and, on the other, there are lasers, which emit light of effectively a single wavelength. In between are fluorescent lighting tubes, giving light which is almost the same mixture as sunlight; tungsten bulbs, which are slightly yellower; candles and oil lamps, which are yellower still; coloured television tubes, and low pressure sodium and neon lights, neon or the vapours of sodium or mercury, which although it consists of only few frequencies, lacks the purity of laser light.

The type of light we want obviously depends on the use for which we need it. Manufacturers of artificial lighting for domestic and industrial use have spent much effort trying both to obtain the exact match of sunlight and to provide 'softer' (less blue) light for social use. For much scientific work, it is necessary to have light which is as nearly pure as possible. Laser light is ideally monochromatic, but expensive. And, of course, there are many situations in which we want light which, although coloured, need not be monochromatic.

The cheapest way to produce coloured light is to remove unwanted wavelengths from white light (see Figure 21). This is usually achieved with a filter, a piece of coloured glass, plastic or gelatine, which absorbs light of some wavelengths but allows other wavelengths to pass through. Other filters can, of course, be improvised. Charles Daubeny, a mid-nineteenth-century professor of chemistry and botany at Oxford, used flattened flasks of port as filters to provide red light for his studies of photosynthesis. Examples of such coloured lights abound (see Chapter 18); traffic signals, Christmas-tree lights, lights on vehicles, lighting in colour printing, photographic darkrooms, theatres and discos. But such filters often let through a band of wavelengths which is too wide for scientific use. Instruments which need more nearly monochromatic light often disperse white light into a spectrum, by means of either a triangular prism or a grating (see page 61).

Once the white light has been fanned out into its components, light of the wavelength needed can be selected by placing a slit, opaque screen over the spectrum in such a position as to block out light of all wavelengths except those in the narrow band which is needed. Since the width and position of the slit can be adjusted, any part of the spectrum can be used, although, of course, when the slit is narrowed, fewer photons pass through it and the beam becomes weaker.

Stronger beams of coloured light may often be obtained from light sources which emit photons only in a limited range of wavelengths, pro-

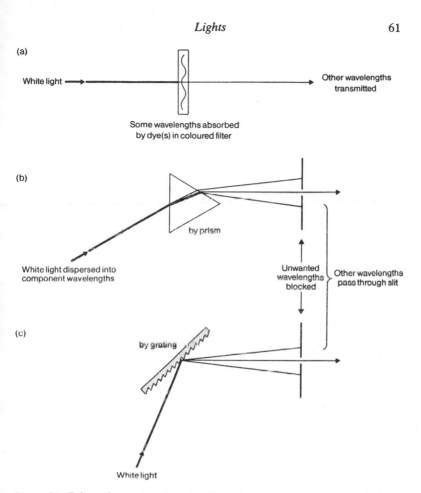

Figure 21. Colours from white light *(a) Absorption of unwanted wavelengths by a filter; (b) and (c) dispersion of wavelengths by (b) a prism and (c) a grating.*

vided, of course, that sources of the right colour can be made. Colour television (Chapter 16) has stimulated the development of coloured phosphors, and strip lights containing neon and other rare gases which emit characteristic colours are much used in advertising (Chapter 18).

We shall be discussing the use of many of their forms of artificial lighting in Parts Five and Six. First, however, we shall look more closely at the appearance of the natural world, by natural light.

PART THREE:
THE NATURAL WORLD

O the opal and the sapphire
Of that wandering western sea.

– THOMAS HARDY

My heart leaps up when I behold
A rainbow in the sky.

– WORDSWORTH

Such thou art, as when
The woodman winding westward up the glen
At wintry dawn, where o'er the sheep-track's maze
The viewless snow-mist weaves a glistening haze,
Sees full before him, gliding without tread,
An image with a glory round its head;
The enamoured rustic worships its fair hues,
Nor knows he makes the shadow, he pursues!

– S. T. COLERIDGE

Why grass is green, or why our blood is red
Are mysteries which none have reach'd into.

– JOHN DONNE

It was the Rainbow gave thee birth
And left thee all her lovely hues.

– W. H. DAVIES, 'The Kingfisher'

Observe the appearance in sunlight of the plumage that rings the neck of a dove and crowns its nape: sometimes it is tinted with the brilliant red of a ruby; at others it is seen from a certain point of view to mingle emerald greens with the blue of the sky.

– LUCRETIUS

Why make so much of fragmentary blue
In here and there a bird, or butterfly,
Or flower, or wearing-stone, or open eye,
When heaven presents in sheets the solid hue?

– ROBERT FROST

Beauty, thou wild fantastic ape
Who dost in ev'ry country change thy shape!
Here black, there brown, here tawny, and there white;
Thou flatt'rer which compli'st with every sight!

– ABRAHAM COWLEY

And, like a lobster boil'd, the morn
From black to red began to turn.

– SAMUEL BUTLER, *Hudibras*

7

AIR AND WATER

On this fine summer morning, the English Channel is almost turquoise, trimmed with white; westwards, successive grassy headlands recede from green, through lovat, to smoky blue. The sky is not unrelievedly azure, for this is England, but the clouds are few, small and mostly white.

Looking westwards in the morning, we should have the sun behind us, so any light we can see must have bounced off something ahead of us. The colours which we see naturally depend on exactly what this 'something' is; on its composition, of course, but also on its size. Most of the 'invisible' particles in our atmosphere are tiny groups containing only two, sometimes three, atoms. They are many trillion, trillion times smaller than the smallest water droplet which we can actually see. But we can see water also as a skein of foam and as the ocean, extending to the horizon. Sunlight, of all visible wavelengths, stimulates oscillations in the tiny, widely separated particles in the air; but these oscillations re-emit mainly high-energy, short-wavelength *blue* light (see Figure 22). Blue light is also preferentially and more strongly scattered by those small particles of bulk matter, such as dust and germs, which, although they contain a large number of atoms, are appreciably

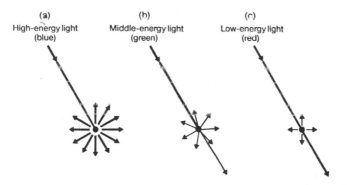

Figure 22. White light and tiny particles *When white light meets particles of size comparable to its wavelength, the high-energy light is preferentially scattered (a). As the energy decreases, the proportion of light transmitted increases (b), so that, for red light (c), very little scattering occurs.*

larger than the wavelength of light. This blue light is responsible for the colour of fine smoke and distant landscapes.

A grassy headland looks green in sunlight because it absorbs most of the visible spectrum except for the green light, which is scattered in all directions from the surface. *En route* to the observer, it is scattered again, by molecules in the air, and so its intensity is slightly decreased. The light which reaches the observer comprises this green light and the predominantly blue light which has been scattered off these same particles. So the green light we see from a distant headland is slightly less intense than from a near one, and it is mixed with considerably more blue (see Figure 23); hence the colour gradation of aerial perspective by which painters endow their landscapes with a sense of depth.

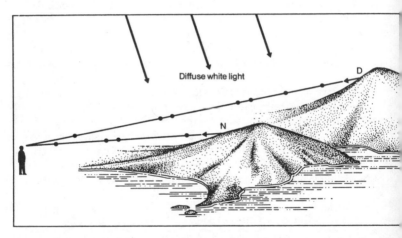

Figure 23. Distant blue *The green light scattered from the distant grassy headland, D, is much the same as that from the nearer one, N. But, on its way to an observer, it meets more particles, which scatter away some green light (cf. Figure 22 (b)) and add some blue light (cf. Figure 22(a)).*

Particles which are large compared to the wavelength of visible light do not scatter colours selectively. When white light falls on water droplets, therefore, that part of the light which is re-emitted, or scattered, is also white (see Figure 24). The smaller the droplets, the greater the surface between them and the air, and the higher the intensity of the scattered white light. Some clouds, which contain a high concentration of droplets, look sculpted rather than wispy. The dazzlingly white, fully illuminated regions of a bank of cumulus may cast shadows on the further parts, which in

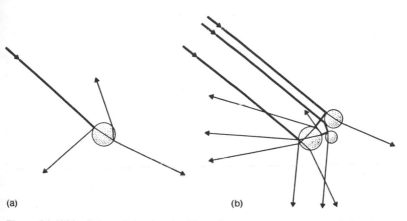

Figure 24. White light and droplets *(a) The reflection and transmission of white light by a single drop of water. (b) With a 'cloud' of three droplets, light emerges almost at random (even ignoring internal reflections),*

contrast seem grey. If a cloud floats very low in the atmosphere, light scattered on to it from the earth's surface may account for an appreciable fraction of its total illumination, and the cloud will then reflect some of the colour of the land, or water, over which it is passing. The low clouds which are all too common over the west coast of Scotland show heather-coloured tints in late summer (see Figure 25).

The sea, of course, reflects the sky. Water is the traditional mirror in mythology, from Greece to China. 'Narcissus joins his image in the lake' and the peasant drowns reaching for the moon. But the sea is not a faithful mirror. It is always darker than the sky because it does not reflect 100 per cent of the light which reaches it; and it is also greener. The light which falls on the sea will be a mixture of white light radiating from the sun, white light scattered from clouds, blue light from the sky and maybe a little light from the land. The light which travels from the sea to the observer comes by two routes. Some of it is reflected from the surface of the sea, and the colour of this is the same as the colour of the light shining on it. Other light may be reflected off the bottom of the sea; the colour of this light obviously depends on the colour of the sea bed. If this were covered with red coral, which absorbs all colours except red, the light reflected from the sea bed would be red. The sea would look magenta, as the observer would receive a mixture of blue and red light. But even if the sea bed were silver sand, the light reflected from the bottom would differ from the surface reflection because the water itself is very weakly coloured. It absorbs red light and so looks bluish green. The depth of the colour naturally increases with the

The Natural World

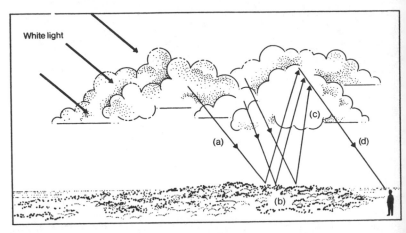

Figure 25. Reflected colour in clouds *White sunlight (a) is scattered from cloud to hilltop. Flowering heather (b) absorbs green light preferentially, so blue and red are scattered preferentially (c) from earth to water droplet and (d) back again. The cloud looks tinged with purple from the heather.*

water's depth; the red component of sunlight is almost totally absorbed by a hundred feet of water. Looking down from a cliff, or from an aeroplane, on to the coastline of a calm sea, we find the shallows charted in pale aquamarine, which darkens with depth, through clear blue-greens to sapphire.

The absorption of red light by water is dramatically illustrated by the Blue Grotto in Capri. This cave is connected to the open sea through a semi-circular hole only about 2.5 metres in diameter. Almost all the light which enters the eye of an observer inside the grotto has been reflected off the very pale sea bed about 7 metres below the surface. As the sunlight therefore passes through at least 14 metres of water, an appreciable proportion of its red component is absorbed. So the water looks gleaming azure (see Figure 26).

As waves break, the water is flung into the air in a myriad of drops; and at each surface between air and water, sunlight is scattered, irrespective of wavelength, as from a cloud. So flying surf looks white, as does the foam at the sea's edge where air is momentarily encapsulated in bubbles of sea.

We return some twelve hours later to see the sun sinking behind black headlands in a blaze of red. The sunlight which reaches us has travelled through a much greater thickness of atmosphere at this late hour than it did around noon. Most of the blue light has been scattered away, as has much of the green and yellow; but the red and orange reaches us (see Figure 27). As

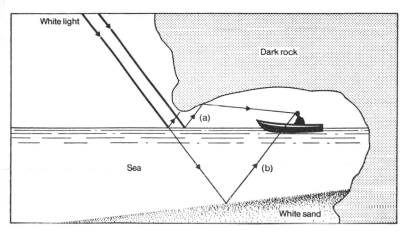

Figure 26. The Blue Grotto *Of the little white light which enters the grotto by path (a), much is absorbed by the rock. Most of the light reaching the boatman has travelled along route (b), passing twice through a 7-metre layer of seawater, which absorbs most of the red light.*

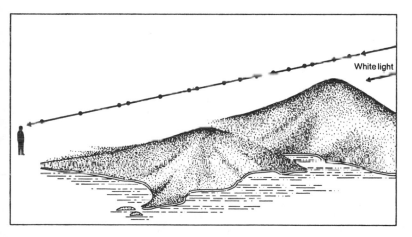

Figure 27. Sunset *No light is scattered towards the observer from the headlands, which now look black. The low rays of the sun encounter many particles before they reach the observer. Almost all of the blue, and much of the green, has been scattered from them; the sun sets golden flame.*

the sky to the east is dark blue, shading to emerald, a little blue light must have got through, and is being scattered back, together with light of the next highest energy, green. In an hour or so the only sunlight reaching us will be that reflected off the moon; and in moonlight we see no colour (see Chapter 11).

Sunsets are much influenced by the number and size of water droplets and solid particles in the air and are particularly spectacular after volcanic explosions which throw large quantities of very fine dust into the atmosphere. The scattering of light can also give rise to a range of less garish colours which can be seen when snorkelling on a sunny day in a calm sea. A swimmer who looks horizontally towards the sun, just under the surface of the water, sees the sea as almost a golden green; much of the blue and violet has been scattered away, and some of the red has been absorbed. An about-turn sometimes produces a transformation scene in which the colour of the sea passes from grape-green and turquoise to a rich, though soft, blue-violet. When we are looking away from the sun, much of the light which reaches us is of short wavelength, scattered from particles which are too small either to sink or to make the sea look cloudy.

Of all the visual beauties of sea and sky, it is surely the rainbow which has most captured man's imagination. Symbol of God's pledge and man's hope, synonym for the infinitely varied richness of colour, no wonder that Wordsworth's heart leapt up when he beheld one. Although the rainbow is only one of a number of celestial arcs, it is perhaps the most spectacular. It is certainly one of the most familiar, at least to those who live in temperate zones with changeable weather. All we need is simultaneous rain and sunshine, with the sun behind the observer and not too high in the sky. Each drop of rain acts as a prism separating the sunlight into colours. If light of one particular wavelength meets a raindrop, some will be reflected back into the air and some will enter the drop; when the light within the drop reaches the surface, some will emerge back into the air and some will be reflected from the inside and travel within the drop until it again meets the surface; here some more will emerge and some be reflected again within the drop. When the light enters or leaves the drop, its direction changes to an extent which depends on its wavelength (see page 23). We can see this prismatic effect quite easily in a dew-drop early on a sunny morning when the light is still low enough to make the dew sparkle. If you focus your attention on a single drop and move your head gradually, you can see in turn each colour of the spectrum (see Figure 28).

When the sun shines on rain, we have a three-dimensional mass of drops, all acting as prisms. After one reflection from an inner surface, each drop will re-radiate blue light at one angle to the sun, red at another and other colours at angles in between. We are not aware of all this, since we see only

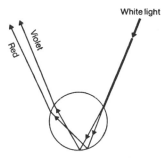

Figure 28. Colours from a dewdrop

the light which falls on the very small area of our own pupils; we have to *scan* the emerging light from a single dew-drop in order to see the different colours. But when we stand and face sunlit rain, we receive blue light from some drops, green from others and so on. These drops which supply an observer with light of a particular wavelength all lie on the curved surface of a cone of which the axis runs through the observer (at the apex) towards the sun. The angle between the axis and the curved surface depends on wavelength; it is 40.6° for blue light, and 42° for red light; other colours lie on cones of intermediate angle. The rainbow is that part of the bases of the set of cones which is visible above the skyline. The blue arc lies on the inside, as the blue cone has the sharpest apex (see Figure 29). As the sun rises in the sky, less of the rainbow protrudes over the horizon, and when the sun is more than 42° above the horizon, no rainbow will be visible (except from an aeroplane). A weaker, outer rainbow may be formed by light which has been reflected twice from the inner surface of drops. The cones of light are less than those formed after one reflection and the angle for red (50.4°) is less than that for blue (53.6°). Thus, although the red band is on the outside of the main rainbow, it is on the inside of the secondary one (see Figure 30). The water drops need not, of course, be rain. Spray from a boat, a waterfall or a lawn sprinkler can also split up sunlight into a spectrum of colour. But spray does not usually cover an area large enough to produce noticeably curved arcs; it can give a vivid strip of rainbow colours, but seldom an actual bow.

The band of colours in a rainbow is not exactly the same as the spectrum of sunlight obtained from a single triangular prism; and, indeed, no two rainbows are identical. For light of only one wavelength, rays from the sun may enter the drop at slightly different angles; so they will travel at slightly different distances within the drop and emerge out of step with each other. As the angle of viewing them is gradually varied, the rays will alternately

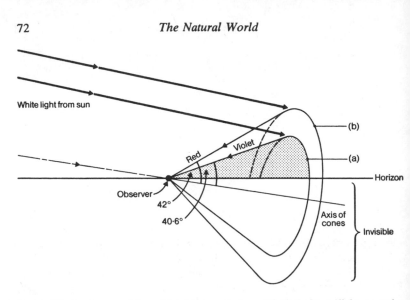

Figure 29. A primary rainbow *The observer receives violet light from all drops on the shaded outer surface (a), and red from those on the unshaded outer surface (b). As the sun gets higher in the sky, that fraction of each cone which lies above the horizon decreases.*

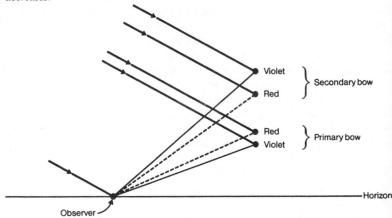

Figure 30. A double rainbow

reinforce and cancel each other, to give a series of concentric bright and dark rings, of that particular colour, superimposed on the main rainbow. The radii of these roundels depend on both the wavelength of the light and the size of the rain drops. From the discussion in Part Four of the effect of mixing lights of two colours, we shall see that this effect accounts for the variation of the brightness, width and even the order of colours in a rainbow. Skilled viewers can use such observations to estimate the size of the drops. When the drops are extremely small, as in fog, the concentric rings of different colours are of closely similar radius. So, except at the extreme edges, the colours overlap. A fog bow is a white band, with a bluish inner edge and a reddish outer one.

Ice crystals, as well as raindrops, can act as prisms and produce circles of bright light. These can be seen around the sun, and quite a distance away from it (the radius of the smallest gives an angle of about 22° at the observer). Although predominantly white, these 'haloes' are coloured at the edges. Ice-crystals, which are usually hexagonal, can also exist as either long needles or flat plates; and since they are randomly orientated, they can produce a complex series of internal reflections. The various emergent beams may result, not only in circles of light, but in a complicated system of bright patches ('mock suns' or 'sun dogs'), ellipses, pillars and even crosses. Interference of light passing through a cloud of droplets or ice crystals accounts for the coloured roundels, or coronae, to be seen around the moon when it is covered by light cloud, or round street lamps on a misty night. The exact colours of the roundels again depend on the size of the crystal or drop. 'Glories', too, are formed by overlapping rings of light of different wavelength. Like rainbows, they are seen when the observer is facing away from the sun; but for a glory to appear, the observer's shadow must fall on a cloud of small droplets of uniform size. This may happen quite often to the airborne, and occasionally to mountain walkers. The light seen by the observer originates directly behind him and is reflected straight back from the rim of the drop. The colours arise through interference between light reflected from different points on the rim. At the centre of the rings, all colours overlap, to give a bright, white patch on which the observer's shadow falls; around it lie a number of brightly coloured rings (see Figures 31 and 32). Since this apparition requires a shadow to be cast on to a cloud, it is likely to be seen by those standing on the brink of chasms and precipices at early morning. If the observer waves his arm, the apparition waves in reply. The experience is unnerving to a cold, tired hillwalker unfamiliar with the phenomenon of the Brockenspectre, or Brocken Bow; and it is all too easy to see why Eastern mystics, thinking Buddha beckoned them, leaped off to join him. The analogous pilot's bow, which surrounds a small shadow of a plane on a cloud below, lacks the supernatural appeal of a halo around one's own head. Since the random scattering of white light from the cloud is likely to be brighter, the colours of the roundel are probably more dilute. A pilot's bow, although a diverting sight, scarcely seems to merit the term 'glory'.

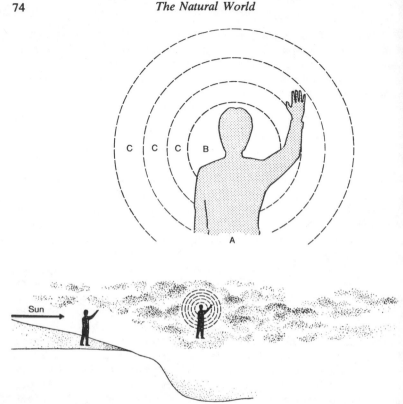

Figure 31. The Brocken spectre *The head of the shadow, A, appears on a bright patch, B, surrounded by coloured rings, C.*

The colours we have so far discussed all originate in sunlight. Energy of a multitude of visible wavelengths arrives at our planet, and a variety of interactions with the atmosphere and with the surface of the earth alters the proportions so that the mixture which is available at any particular point may not be 'white'. Sometimes, however, earth may be lit from other sources. During a thunderstorm, clouds become electrically charged through friction; electrons are transferred from one mass of droplets, hailstones or ice crystals to another in much the same way as nylon clothing becomes electrically charged in a rotary drier. When the excess, or deficiency, of electrons is large enough, there will be an electric discharge in which the electrical balance will be restored. There is a massive increase in temperature (up to 25,000°C). The air expands violently, and there is much reshuffling of particles and emission of energy as the air returns to its

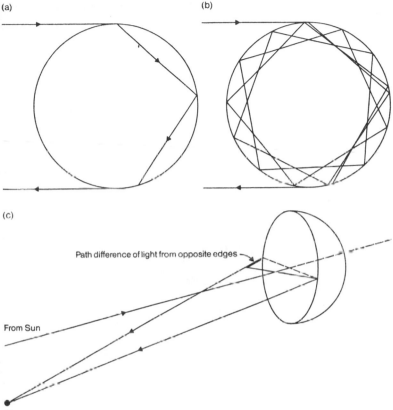

Figure 32. How a glory occurs *Light entering a drop at its edge is reflected back in the same direction, after reflection (a) once, or (b) fifteen times from the inner surface. For drops straight ahead of the observer, all points on the edge are the same distance away, and no interference occurs. But from drops at an angle (c), the outer edge is farther from the observer, and interference may occur, giving bright and dark rings, at different angles for different colours. (Adapted, with permission, from H. C. Bryant and N. Jarmie,* Scientific American, *vol. 231, July 1974, p. 60.)*

original, stable state. In the visual region, energy of over two hundred different wavelengths is emitted, much of it in the high-energy region. So lightning, like the sparks from electric trains and faulty electrical connections, is white, with a tinge of bluish-green. (But, seen from an aeroplane flying above a distant storm, lightning can have a faintly pink look, presumably because, if the flash occurs below the plane, some of the higher-energy light is scattered back towards the ground.)

A somewhat similar electrical discharge occurs during the aurora, which is powered by the bombardment of air by electrons and other small, electrically charged particles which stream in from the sun, sometimes in great quantity and at great speed (about ten million per square centimetre per second at several thousand kilometres per second). The particles are attracted to the earth's magnetic poles, and it is in these regions that auroral displays are the most spectacular. The commonest colour, yellowish-green, is caused by changes involving oxygen atoms at low pressure. Higher in the sky, and therefore visible at lower latitudes, the changes involve oxygen atoms at still lower pressure, and the light emitted is red. Rarer bluish, or red-edged green, displays are produced by pairs of oxygen or nitrogen atoms which have been excited, or have lost one or more of their electrons.

8

EARTH AND FIRE

We are by the sea again, but now restrict our attention to the beach, which is part sand, part shingle, punctuated with rocks. Along this stretch of the coast, the colours are not striking; the rocks are brown and the shingle seems, at first glance, a mottled, mealy grey. A closer look resolves the shingle into lacklustre pebbles of various pale browns and greys, mixed, perhaps, with some white ones, and a few which are dull black or purple. At the shoreline, however, the scene comes to life. Reflection from the wet surfaces gives highlights and the colours themselves become deeper and richer. The sand turns from matt fawn to gleaming ochre, and the pebbles look smooth as polished marble, and appear in as wide a variety of colours.

Few rocks are as dazzling white as Parian marble or the chalk cliffs of Dover. But many are coloured only because they contain traces of some contaminant in minerals, such as quartz, limestone or gypsum, which, if pure, would be colourless. By far the most common of these impurities is iron. Usually, each iron atom has shed three of its outer electrons on to neighbouring groups, but, by absorbing some visible light, it can temporarily reannex one from an oxygen atom of any silicate, carbonate or sulphate group which may be close by. This transfer of electrons from oxygen to iron accounts for all the ochres, rusts, browns and reds in our rocks, sand and buildings. The colour is very sensitive to the structure and water content of the mineral, as illustrated by the change in colours when clay is fired. The blue-grey streaks in some clays and cream limestones are also caused by iron atoms, but by those from which only two electrons are lost. Prolonged exposure to air, or heating to form bricks, make the iron lose a third electron and change to its more familiar ochre form.

Since pebbles, and sand, are merely moveable and hence worn fragments of rock, why do they look so much paler? A grain of sand, being so small, has a very large surface compared to its volume, and so it scatters a high proportion of the white light which falls on it. The ochre colours of any iron it may contain are therefore much diluted, and the sand looks paler than the rocks from which it was formed. The finer the sand, the whiter it looks (and the same effect may be obtained by crushing demerara sugar or coloured glass). The paleness of pebbles arises in the same way: their surfaces are scratched and pitted from movements against other stones and so act as effective scatterers of light, regardless of its direction or wavelength. Hence, their pale, dull look.

Rocks, pebbles and sand all show darker, richer colours when they are wet, though the effect is less noticeable for rock, particularly if it has a fairly smooth surface. When light passes from one transparent material into another, that proportion which is scattered depends on the relative ease with which light travels through them. If light travels at almost the same speed in the two substances, there is less scattering than if the speed of light changes appreciably at the boundary. Since light is slowed down more on passing from air to stone than from water to stone, sand and worn pebbles scatter light more effectively if they are dry. When light travels from water to sand or pebbles, less of it is scattered and more enters the stone. Here light of some wavelengths is absorbed and that of others transmitted or reflected back to the observer, so the light which reaches us is less diluted and the colours darker and richer. The greater the scattering power of the dry material, the more marked is the effect of the water; new, shiny plastic, which reflects (rather than scatters) most of the light falling on it, is almost unchanged in appearance when it is immersed in water.

The wide range of pebble colours revealed by the sea is not entirely because of the presence of iron. Other metals also contribute, but the colours they generate arise from a change in orientation of electrons within a particular atom rather than through a temporary electron transfer from oxygen atom to the iron. As such changes are of low theoretical probability, colour is generated only when appreciable concentrations of the metal are present (see page 40). The pink, or purple, of rose or amethyst quartz, is caused by manganese, while minerals, such as malachite, which contain copper are often green or blue. In some minerals, the metal ion is surrounded by atoms of sulphur rather than of oxygen. Transfer of electrons from sulphur atoms to the metal is fairly easy, and is accompanied by intense absorption of light of a wide range of wavelengths. Such minerals are black.

Many of the colours of gems and semi-precious stones, as of more mundane rocks, are caused by the contamination, by traces of metal, of basically colourless, oxygen-containing materials. Ruby and emerald both owe their colour to chromium. The dramatic difference between the two is owed merely to the fact that the enveloping oxygen structure in ruby, but not in emerald, is under considerable internal stress. Sapphire is structurally similar to ruby, but contains iron and titanium instead of chromium. The colour of deep-green jade, like that of emerald, is caused by chromium in an unstressed oxygen environment. Paler green jade is coloured by iron atoms which have each lost two electrons, unlike the ochre colours which often arise when three electrons are removed from the iron. Turquoise, as can be guessed from its similarity to copper sulphate crystals and to weathered copper domes, owes its colour to copper.

Metal ions are not, however, the only components of rock which give rise to colour. Some minerals contain the bulky disulphide group clouds whose electron clouds are so diffuse that they overlap with those adjacent groups. The shared clouds which are formed are midway between 'clasp' and 'treacle' in type. They absorb visible light and give rise to a variety of colours: royal blue in lapis lazuli, black in marcasite, metallic golden yellow in iron pyrites (fool's gold). Graphite, although a form of non-metal carbon, looks almost as shiny as a metal. The atoms form layers, and on either side of each layer is a continuous sheet of electron cloud, like a multi-decker treacle sandwich. And it is these layers of electron cloud which give graphite its metallic lustre (see Figure 33).

Crystals of rubies, sapphires and other gems can now be grown artificially, and it is found that the colours are often improved by heat treatment.

(a)

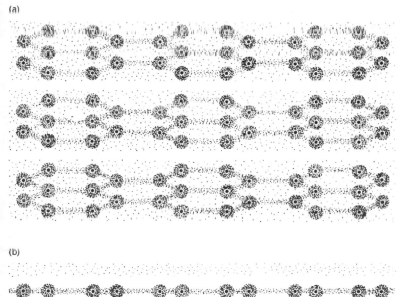

(b)

Figure 33. Shiny graphite *The structure of graphite, illustrating (a) three layers of carbon atoms, and (b) end-on view of a single layer, showing electron cloud above and below the sheet of atoms.*

The colours of natural gems, too, can be permanently changed in this way. Zircon (zirconium silicate), for example, can be changed from brown to red, yellow, green, blue or colourless. Presumably the change in temperature either alters the detailed arrangement of the oxygen atoms around the zirconium ion or changes the number of electrons associated with it. But these improved colours may fade on prolonged exposure to sunlight, and so heat-treated gems are seldom displayed in jewellers' windows.

The colour of the semi-precious gem alexandrite is very sensitive to the light which falls on it. The mineral itself absorbs blue and yellow light. In daylight, it transmits much green light, together with some red, and so looks a fresh, yellowish green. Since artificial light contains a lower proportion of high-energy and middle-energy light than does daylight, so does the light transmitted by the alexandrite. In tungsten lighting, the gem transmits roughly equal amounts of green and red, so looks yellow. By candlelight, which is of still lower energy (see page 56), the alexandrite looks red.

But colour, as we have seen, is not always caused by absorption of light. The coloured sparkle of diamond is like that of a particularly impressive dewdrop or chandelier. When white light passes from air into diamond, the colours are spread out more effectively than they are when they enter glass; and a much higher proportion of the light is reflected internally from the back of a diamond than from the rear surface of a drop of water or a glass prism; and on reflection, the spread of colours is amplified. Diamonds are cut so that the light is reflected at the inner surfaces of a number of facets: the various colours emerge from the front of the gem at markedly different angles. When the wearer, or the observer, moves slightly, the light which enters the eye is first of one almost pure colour, then of another. Modern imitations may disperse and reflect light equally impressively. Their sparkle rivals that of a real diamond, though they lack its extreme hardness.

Some of the most beautiful colours of the mineral world are caused by optical interference. Precious opal, like soap solution, absorbs little visible light; it is made up almost entirely of silica and water. But part of the structure consists of regularly stacked spheres, and reflections from neighbouring groups of these may interfere. Depending on the angle of viewing, light of one or more wavelengths is cancelled out. The opal acts as a diffraction grating (see page 49). Legend maintains that the exact colour of an opal is sensitive to the well-being of the wearer, and it is indeed possible that the spacing between spheres varies with changes in temperature and in the humidity of the skin. The mineral labradorite shows similar iridescence, usually in the colour range from royal blue to kingfisher, green and gold. These peacock colours are superimposed on a background of dull grey, which is caused by thin sheets of oxides of iron or titanium embedded in a colourless felspar mineral. The iridescence arises through the interference of light reflected from different layers of oxide, the exact colour depending on the spacing and orientation of the layers and on the viewing angle. The less exotic iridescence observed by Goethe on 'the surface of stagnant water, especially if impregnated with iron', interference also arises from layers of iron oxide.

We have seen that many solids change colour when they are heated. The change may be permanent; clays and earth pigments, such as raw sienna, lose water on heating and acquire a much redder hue. Their structure is permanently changed, and their original ochre colour does not return if they are wetted. Some effects of heating can, however, be reversed. We may remember from our school days how blue copper sulphate crystals lose water and crumble to a white powder when heated, the blue colour being restored when water is added. This colour change, too, arises from the change in the environment of the metal, and hence in the exact arrangement of its individual electron cloud. A similar colour change has been used as a 'dampness' indicator for drying agents and 'weather pictures'. The pale pink colour, indicative of dampness, arises from cobalt surrounded octahedrally (by four particles of water and two chloride groups), whereas the brighter blue, dry-weather colour is produced when the cobalt is in the centre of four tetrahedrally disposed chloride groups (see Figure 34). The blue is more intense than the pink because changes within a tetrahedron are more probable than those within an octahedron. Environment changes also cause purple crystals of chrome alum to turn green as they dissolve in water (c.f. page 78), and the 'Cambridge' blue of a copper-sulphate solution to darken to 'Oxford' blue if ammonia is added.

Some reversible colour changes can be produced merely by raising the temperature. Another survival from school chemistry is 'zinc oxide; yellow when hot, white when cold'. The hot form has sufficient energy to be able to undergo small changes in structure on the absorption of visible light.

Very hot solids (and liquids) glow; they have so much energy that they emit not only infra-red radiation but also low-energy visual radiation and become 'red-hot'. If their temperature is raised still further, they may become 'white-hot', giving out energy over the whole visible range.

Many substances, when heated in the presence of air, combine with oxygen and change colour. A bright copper coin merely acquires a black coating of copper oxide, but for many substances, 'oxidation' is accompanied by burning. The light emitted by the flames often shows beautiful colours which depend on the substance which is being burned and on the rate of supply of both fuel and air to the different regions of the flame. We are so familiar with domestic flames from matches, candles, cooking gas and fires that it may seem surprising that many of the changes which take place within them are extremely complicated and by no means fully understood.

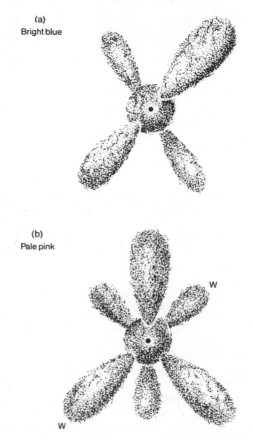

(a)
Bright blue

(b)
Pale pink

W

W

Figure 34. Fine blue, damp pink *Clouds of electrons around a central cobalt nucleus in combination with chloride groups (a) when dry, and (b) after absorption of a little water, W.*

If we look at a steady candle flame, we can see three main regions: a transparent blue base, a greyish semi-transparent central cone and an opaque, bright-yellow outer cone. When the candle is lit, the wax liquefies, rises up the wick and vaporizes. Oxygen diffuses into the gaseous wax and combines with it, eventually producing carbon dioxide and steam while giving out considerable heat. At the high temperature produced, the wax breaks down into very small, highly excited groups of atoms (such as 'dicarbon' and 'carbon-plus-hydrogen') which, at the base of the flame, rid themselves of some of their energy by emitting blue light. Higher up, the carbon atoms may join together and form relatively large conglomerates, or soot particles, which become 'yellow-hot' and glow, forming the bright, opaque outer cone of the flame.

Flames produced when other substances burn may naturally contain other small, excited groups of atoms which will emit light of other wavelengths. Coke, for example, burns with a clear blue flame, produced by excited oxides of carbon. The flames of domestic cookers and heaters differ from those of candles or fires because air does not just seep into the hot gaseous fuel but is premixed into the unignited gas. In coal gas, the colour of the flame varies markedly with the proportion of air in the mixture. A low ratio of air to fuel produces a yellow flame similar to that of a candle, while a high proportion of air to gas gives a hotter, clearer blue flame with a turquoise central cone, containing oxygen-plus-hydrogen groups. Natural gas is chemically more akin to candle wax and, when mixed generously with air, burns with a blue flame composed of an inner cone similar to that at the base of a candle flame, a paler, greener middle cone and a more purple outer cone.

Flames, like solids, can be strongly coloured by the presence of very small amounts of impurities. A crystal of table salt will colour the flame of a gas stove bright yellow. Pale mauve flames sometimes flicker over a particularly hot bonfire, and if any fine copper wire finds its way into a fire, it colours the flames emerald green. These colours are owed to individual atoms which have gained so much energy from the fire that one or more of their outer electrons has been promoted to regions even further from the centre of the atom. As we saw on page 56, the colour of the light emitted when the atoms revert to the state of lowest energy is characteristic of that particular type of atom. The yellow flame of table salt is exactly the same as that produced by washing soda, caustic soda or sodium bicarbonate and is caused by the presence of sodium.

Flame colours have long been used to identify chemical elements: a deep green flame is characteristic of copper; and pale mauve which is often seen over a wood fire, is caused by potassium, which is absorbed by plants from the soil. Nowadays, the intensity of flame colour can be used to measure both the amount of the element which is present and the temperature of the flame. Atomic flame colours are exploited more frivolously by manufacturers of fireworks, who have access also to less familiar substances. The heads of 'Bengal matches', for example, contain either lithium, which produces a rich crimson flame, or barium, which gives a limegreen one. Relatively few elements, however, produce coloured flames, which accounts for the rather restricted palette available to the pyrotechnist.

9

VEGETABLE COLOURS

Since only a small fraction of dry land is covered with bare rock, sand or snow, or with buildings, minerals make a limited contribution to the colour of the earth's surface. Seen from the air, most of the land, as any air traveller can testify, is covered with vegetation: forest, grassland, crops or scrub. During the growing season, leaves account for almost all the surface area of most types of plant. The landscape looks mainly green while the leaves are young, though the colours may change through yellow and red to brown as the year proceeds. But it is not solely the leaves which impart their colours to the landscape. A bare forest shows a range of subtle browns and the heather-clad hills of Scotland are smoky purple. Cultivation has brought us orchards of fruit blossom and fields of spring bulbs. European farmland is studded with patches of dusky-blue lucerne and acid-yellow rape. Few British wild flowers, other than heather, monopolize enough expanse to determine the colour of a distant landscape, but several provide more restricted patches of showy colour. Newly cut roadsides can be pink with willowherb or scarlet with poppies, and the floor of a woodland can look almost solidly coloured, white or blue, with anemones or bluebells.

Focusing more closely on to individual plants, we can see the immense variety in the colour of the flowers, and also, perhaps surprisingly, of the leaves. Cabbages may be greenish cream, bottle green, smoky turquoise, purplish blue, rich claret, or glossy black as an aubergine. The emergent shoots of daffodils are pale sulphur yellow, while bursting leaf-buds of peony shine like old mahogany.

Much of the beauty of vegetation in a landscape arises from its movement. If we walk through a beechwood towards the sun, we see light which has passed through some leaves and bounced off others, some of them shaded by their neighbours. So the light entering our eyes will vary greatly in colour and in brightness, changing with the breeze; and the forest will look quite different if we turn away from the sun. We can see a similar play of light and shade, yellow-green and blue-green, from the stripes on a freshly mown lawn. The mower bends alternate bands towards and away from the viewer; those bent away from him reflect more light when observed against the sun and so look a lighter and yellower green. And, when wind ripples a cornfield, the colour is as of shimmering velvet, and for the same reason (see page 47).

Almost all plant colours are produced by substances which are themselves coloured; colours which arise by interference or scattering are rare. Vegetable colours, like other coloured materials, absorb light of some, but not all, energies in the visible range (see page 37). Green leaves, for example, contain chlorophyll, a substance which absorbs all colours except green. Its structure is complex, and somewhat like a badly made lace mat, ordered in the middle, but ragged around the edges (see Figure 35). At its centre is a single magnesium atom, stripped of its two outer electrons. Around it is a square of four nitrogen atoms enclosed in a flat, openwork arrangement of twenty carbon atoms. Outside this regular structure are a number of short, branched 'arms' consisting of different small groups of atoms, and one 'tail', a chain of about twenty carbon atoms. In this assembly, the most loosely bound electrons are those which lie above or below the 'lace mat' or above and below its 'tail'. These rather mobile electrons can absorb rather small amounts of energy, in the visible and ultra-violet regions of the spectrum, but they readily revert to the more stable state by transferring their extra energy to some neighbouring substance.

It is this power of chlorophyll to act as an energy carrier that enables life to exist on earth. Green plants absorb energy from sunlight and use it to drive the process of photosynthesis in which complex energy-rich sugars are built up from the much simpler water and carbon dioxide which the plant takes in from the soil and from the air. Not only does photosynthesis trap energy from sunlight and store it in food, but, as a by-product, it produces oxygen, which almost all living things need if they are to make use of energy stored in food. So to chlorophyll we owe not only our food but also the means by which we use it.

Also present in almost all green plants are members of a large group of yellow, orange and red substances which are closely related to the 'tail' in chlorophyll. This electron-rich chain forms the backbone of the assembly (see Figure 36). Attached to each end is a ring of six atoms, and protruding from it are a number of small side-groups. As in chlorophyll, there are electrons on either side of the chain, and these can absorb energy in the visual region, giving rise to the colour of carrots (whence the group name of 'carotenoids'), tomatoes, sweet corn and marigolds. Carotenoids are also present in the leaves of most plants. Brown seaweeds owe their colour to absorption by both chlorophyll and carotenoids, but in flowering plants the green of chlorophyll usually swamps the colours of other substances. The yellow pigments show clearly in autumn, when the chlorophyll decays and loses its colour.

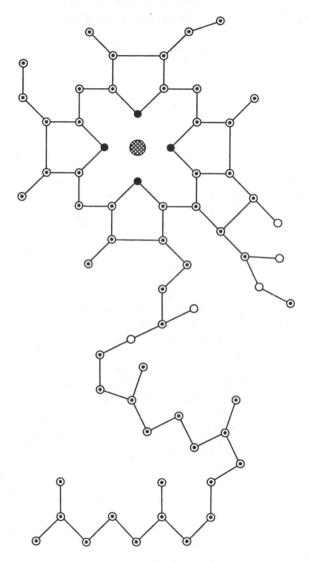

Figure 35. Leaf green '*Lace-mat and tail*' skeleton structure of chlorophyll – α using the same symbols as in Figure 12 (page 41). The central metal, represented by ●, is magnesium.

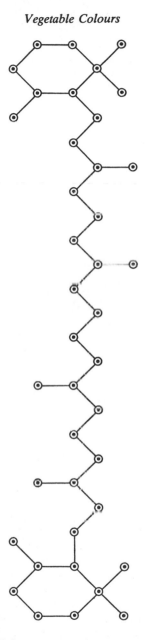

Figure 36. Carrot orange *Skeleton structure of β–carotene, showing carbon nuclei* ⊙.

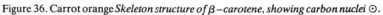

Another large group of vegetable colours, the flavonoids, produces the tints of autumn leaves, the purple shades of many new shoots and the varied hues of a large number of flowers. The parent substance, flavone, has a basic skeleton of three rings. Above and below its main skeleton it, too, has clouds of loosely bound electrons which enable it to absorb visible light (see Figure 37). Variety of colour is provided by small branches of different groups of atoms around the rings by sugars and other substances combined

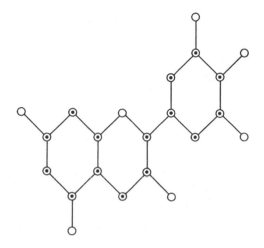

Figure 37. Delphinium blue Skeleton structure of the anthocyanin delphinidin, showing nuclei of carbon ⊙ and oxygen ○.

with the flavonoids. Many of the colours vary with acidity. One group is red in acidic solutions, mauve in neutral media and blue in alkalis. Substances of this type act as natural indications of acidity (see page 203). They are present in cornflowers, in which the pink variety has more acidic sap than the common blue form; in milkwort, which is pink on acidic, peaty soil and blue on chalk downs; and in milkwort and some types of forget-me-not, in which the acidity of the flower increases with its age. Another group of flavonoids is colourless in acidic media but turns yellow when alkaline. The two classes are named after their alkaline colours, the red-to-blue ones being called *anthocyanins* and the colourless-to-yellow ones *anthoxanthins*. Members of the two classes often occur in the same plant, and as each component will probably change colour at a different acidity, the final colour will show a complex dependence on acidity. The juice of pickled red cabbage, for example, is red only if it is acidic; but if ammonia or washing soda is added, the acid is neutralized, and so the anthocyanins turn blue and the anthoxanthins turn yellow; thus the juice changes through various shades of purple, blue and turquoise to green (see Figure 38).

The formation of flavonoids may be affected both by external conditions and by the health of the plant. The red anthocyanins which are responsible for many of the flaming colours of autumn leaves are formed most effectively in bright light at low temperatures, and so autumn tints are brightest when the days are clear and crisp. Some plants also produce red anthoxanthins if they suffer injury or mineral deficiencies. On the other hand, the

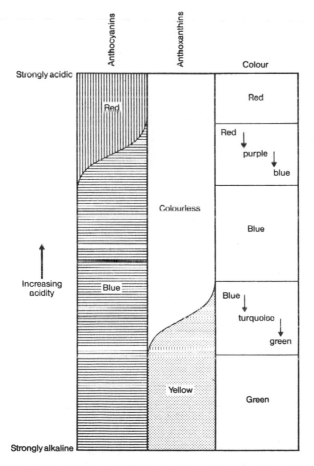

Figure 38. The colourful juice of pickled red cabbage *If ammonia is gradually added to the juice, the colour changes down the sequence on the right.*

natural red or pink colour of some flowers, such as stocks and sweet peas, can be destroyed by a virus which prevents the formation of anthocyanins.

Since many flower petals contain several carotenoids as well as anthocyanins and anthoxanthins, it is easy to understand the enormous variety of colours exhibited by flowering plants. One can sympathize with the view, implied by many a Victorian hymn-writer, that God painted the flowers for the delight of man. However, since most flowering plants which are pollinated by birds and insects (rather than by wind) have showy flowers, it seems likely that these colours also confer some advantages on the plant. The way in which plant colours have evolved has been the subject of much discussion, but it seems clear that coloured petals do indeed attract pollinators and that guidelines may lure them towards the nectar (and hence the pollen). Some members of the orchid family have flowers which closely resemble, in both shape and colour, the female of a species of insect. They are pollinated by the male of that species, who lands on the flower and attempts to copulate with it. The effectiveness of these devices depends, of course, on the ability of the pollinators to distinguish between different colours; but we shall defer discussion of colour vision, in animals and in man, until after we have looked at some of the colours of animals themselves.

10

THE COLOURS OF ANIMALS

Most animals, unlike most plants, are mobile, and so cannot become a permanent part of the landscape. Although coral reefs persist for centuries, and certain rocks are almost permanently teeming with sea birds, most animals are but fleeting occupants of a scene. They produce minimal visual impact on the landscape partly because the larger animals are usually drab, and partly because even a large group of large animals is dwarfed by the landscape it inhabits.

The subdued colouring of many of the larger mammals is not, however, typical of the animal world as a whole. The colours flaunted by many insects, birds and fish rival the brilliance of any florist's or greengrocer's. But whereas plant colours are produced almost entirely by absorption, animals achieve this brilliance by a wider variety of methods.

As an unexotic illustration, we may envisage a fishmonger's stall. In attendance are a tabby cat and her ginger-and-white kitten. The kitten has a pink nose, set in a splodge of white fur, and its eyes are still blue. Despite the variety of colours, the tabby and ginger fur contains only a single class of pigments, called *melanins*, which are responsible for all the brown, black, reddish and golden colours of mammalian fur and hair (the sole exception being red human hair). The structural unit of all the melanins is similar to the regular, lace-mat part of the chlorophyll skeleton, stripped of its central metal atom and neatly quartered. White fur contains no colouring matter, but many tiny pockets of air. White light is scattered from the transparent walls of these air cells in just the same way as it is from a foam of tiny bubbles (see page 68).

The skin on the kitten's nose is also colourless. The kitten's nose looks pink because tiny capillaries carrying red blood lie just below the surface. Haemoglobin, which gives blood its colour, has a lace-mat skeleton similar to that in chlorophyll (see Figure 12); but the central metal atom is iron rather than magnesium. (Hence the use of iron pills in the treatment of anaemia.) Melanin, with its 'quarter-lace-mat' skeleton, is probably formed by the breakdown of haemoglobin.

Surprising though it may seem, there is no blue pigment in a kitten's eyes; only a deposit of minute white flecks backed by a layer of melanin. The particles are small enough to scatter high-energy light more effectively than light of longer wavelengths. So light which reaches the observer from kittens' eyes, like that from fine smoke or from the sky, is predominantly blue. Little white light is scattered from the black background.

As the kitten gets older, a yellow layer will form over the white granules; and so its eyes will become green, like its mother's. The yellow layer contains carotenoids, which are present in most animals as in most plants. Some feline aristocrats have heavily pigmented layers, producing the copper eyes of Blue Persians. Siamese, on the other hand, have no yellow layer and their eyes remain blue; but the white granules gradually enlarge and scatter blue less effectively, and so the eyes become paler with age.

Our kitten is investigating the fish: gleamingly iridescent mackerel, vermilion boiled lobster and drab plaice. The mackerel is a particularly lovely example of the silvery-green shimmer caused by tiny corrugations in the (colourless) scales of many types of fish. When the light which falls on them bounces off the surface, there is some interference between light which is reflected from adjacent ridges. The gap between the striations of mackerel scales is about the wavelength of red light. Much of the red component of white light is eliminated, and so the light which reaches the observer is bluish green. The intensity of the colour varies with the density of the melanin markings in the skin beneath the scales. In rainbow trout, the ridges are of more varied spacing, so that we can see a range of colours, changing with the viewing angle.

The bright red of a lobster, like the similar colour of a tomato, is owed to carotenoids. But lobsters are red only when boiled; living ones are blue-black. Crabs, prawns and shrimps also become much pinker when they are boiled. In the living shell-fish, the carotenoids are joined to proteins which modify their colour, dramatically in the case of the lobster. Boiling the lobster separates the protein from the carotenoid, and we see the much brighter colour of the free pigment.

The white parts of the shells of lobster, and of mollusc shellfish such as oysters, mussels and scallops, is caused by a scattering of light off-white particles, often of chalk. (These particles are, of course, much larger than those in kittens' eyes, otherwise shells would look blue.) The inner shells (and pearls) of some molluscs are iridescent, showing colours which, like those of fish scales, are caused by the interference of light reflected from tiny ridges in the shell. Mother-of-pearl colours are often rather pale, because so much white light is scattered by the white material of the shell. The pinks and oranges of the flesh of many shell-fish, whether of horny-shelled species, such as crab or lobster, or hard-shelled ones, like mussel and scallop, are again due to carotenoids.

Some of the fish on the slab will look much less exotic. Plaice and sole, for example, look mottled khaki. The browns and blacks are caused by melanins, as in a tabby cat; but the surfaces of fish, unlike those of mammals, often also contain carotenoids, which provide a background, or spots, of yellow or orange. We shall see later (page 96) that this combination of black and yellow pigmentation can be of great value to a number of reptiles and amphibians as well as to fish.

If the fishmonger sells Mediterranean sea-food, he may display octopus or squid in its ink, which, again, is melanin. He may even have sea-urchins in their shells, which may be orange, red, purple or green, and are coloured by a class of pigments with much smaller skeletons than those of the carotenoids or melanins. Sea-urchin colours are produced by pigments somewhat similar to the flavonoids; and these colours, too, vary with acidity, which may account for the range of colour seen on a single sea-urchin shell. These particular pigments are found in no other animals, except in the pink or purple bones of sea otters, who eat a lot of sea-urchins.

And what of the fishmonger himself? Unless he is an albino, he will have deposits of melanin in his hair and under his skin. The density of the melanin will depend largely on his race, though in some races it varies greatly with the genetic make-up of the individual. The density of melanin in the skin also varies with exposure to sunlight, and, if the skin is fair, the red blood in the surface capillaries will contribute to its colour. Human eyes, like cats' eyes, contain minute white granules which scatter blue preferentially, and may be covered with a pigmented layer. So their colour may range from blue through grey, hazel, chestnut to dark brown or nearly black, depending on whether there is no layer of colour, a lightly pigmented layer of melanin, or a very dense layer. So the structures which give rise to the colour of the fishmonger are much like those responsible for the colour of his cat; unless he is ginger haired. Red human hair, unlike the red hair of any other mammal, is produced by an unusual pigment which contains iron.

The colours of many other creatures are produced by means very like those we have just described. The opaque white of feathers, like that of fur, arises from the scattering of light by the walls of air-pockets. That in the wings of some butterflies is produced in the same way, but the white in other species of butterfly and in some tropical fish is caused by the presence of white granules.

Many birds and insects have iridescent colours caused by ridges either in a surface layer of transparent material or in an underlying melanin layer. Particularly exotic are peacocks and humming birds, but there are many less flamboyant examples among British wild birds: wood pigeon, starling, pheasant, mallard. On a smaller scale, but no less exquisite, are dragonflies, bluebottles and many beetles, whose wing-cases may be iridescent rusty gold or peacock blue against a jet background; and it is all done by ridges

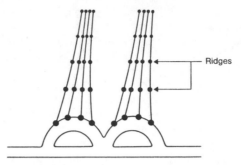

Figure 39. Butterfly wings *Section through two scales on a typical wing of an iridescent butterfly.*

(see Figure 39). The striated wing-cases of one South American beetle produce a different effect; they act as diffraction gratings (see page 49), so that when the insect is in direct sunlight, it appears to emit tiny rainbows (Figure 40).

Although no blue pigment has been found in any vertebrate, blue is by no means an uncommon colour among the higher animals. In every example, from the feathers of the jay, blue-tit, kingfisher and hyacine macaw to many tropical fish, the tail of the blue-tailed skink, the neck of a turkey and the rump of a baboon, the blue colour arises from scattering, just as in the eyes of kittens. The effect is often enhanced by an underlying layer of melanin, and may be modified by a transparent, pigmented overlay (see Figure 41). For example, many green lizards, snakes and birds contain no green pigment; their colour arises from a transparent yellow layer above the granules which scatter the blue light. And, in the violet feathers of some tropical birds, the granular layer is overlaid with red. The red film can be scraped off to reveal a blue feather, which, if crushed, gives a black powder, as the white granules are swamped by the melanin from the backing. Purple regions on baboon rumps and turkey necks are caused by the combination of scattering by a granular layer and absorption by haemoglobin in the capillaries.

Among the light-absorbing substances found in animals, carotenoids supply many of the yellows and reds, as in starfish, salamanders and corals; red birds, such as flamingoes and scarlet ibis, lose their brilliance when in captivity, unless they have a carotene-rich diet. Many of the other pigments are formed by the breakdown of haemoglobin. The black and brown melanins are a particularly important group. The pigments in bile are often 'opened-up' relatives of the lace mat in haemoglobin and give rise to the blue colours of birds' eggs, some blue butterflies and blue coral.

(a)   (b)

(c)

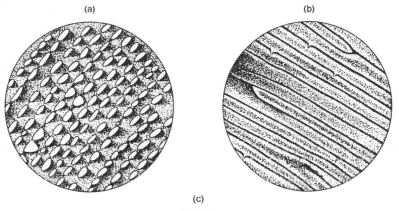

Figure 40. Insect iridescence *Striations which give interference colours, from (a) scarab beetle, (b) rove beetle, (c) mutillid wasp. (Reproduced, with permission, from R. L. Gregory and E. H. Gombrich,* Illusion in Nature and Art, *Duckworth, London, 1973.)*

Black centre — — — Transparent coating

— Cells containing white granules

Figure 41. Blue birds *Section through a branch of a typical blue feather.*

Some types of colour are restricted to a few species or to one alone. The intense scarlet of the primary feathers of Hartlaub's turaco is owed to a rare copper-containing pigment which can easily be converted into a dark-green form which colours the secondaries. Simpler substances, which resemble some plant pigments, are concentrated in certain insects, who presumably extract them from their food plants. Cochineal is a familiar example. Closely related to it is kermesic acid, which was much used as a red dye in the classical world. Tyrian purple, another dye famous from ancient times, is produced by the action of light on a colourless substance secreted by several types of snail.

Animals do not always stay the same colour throughout their lives; changes may occur with age, season, environment and mood. The colours of many young animals differ from that of the adults: chicks are often striped, fledgling birds are almost always drabber than their parents, kittens' eyes turn green with maturity, lion cubs lose their spots and tail-rings, human babies darken. At puberty, human females become lighter and males darker, even in parts not exposed to the sun (cf. page 215). Towards old age, hair grows without melanin in humans and around the muzzles of dogs. Some birds have markedly different winter and summer plumage, and in winter, arctic mammals, such as the fox, hare and stoat, turn white, while some species of deer which live among deciduous trees lose their spots.

A surprising number of animals can change colour to match their surroundings: the fame of the chameleon is only because of the range and rapidity of its reaction, and even in these respects it has no monopoly. Some fish change colour very slowly and semi-permanently according to the illumination of their surroundings, by increasing or decreasing the number of cells which contain pigment; and plaice may slowly grow spots if they move from a sandy part of the sea bed to a pebbly one. Certain spiders are able to change colour over a period of two to three weeks and assume colours similar to the pink, white or yellow flower-petals among which they lurk. It has been recommended that minnows which are to be used as bait in dark waters be 'conditioned' by keeping them in a tin with white sides; they become pale, and more conspicuous to predators. Human beings, on the other hand, may become darker when exposed to sunlight because the melanin deposits become denser; in fair-skinned races, this causes tanning or freckling, depending on how evenly the pigmented cells are distributed.

Rapid, or even instantaneous, colour changes occur in fish, frogs, lizards, squids, shrimps, prawns and a few insects. It is usually accomplished by expansion and contraction of a cell which contains melanin, so that the colour may either be spread out over a large area to give a dark colour, or be concentrated into a dot, allowing the underlying pigmentation to show through (see Figure 42). Many of these changes, which result in the animal matching its surroundings, can cover a wide span of colour; frogs and lizards may turn from emerald green to black.

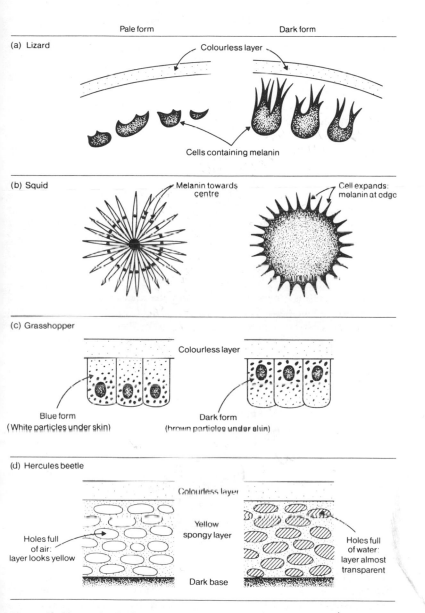

Figure 42. How animals change colour *(a) Section through skin of a green lizard, showing expansion of cells which contain melanin. (b) Isolated cells containing melanin in the squid. (c) Section through skin of blue grasshopper, showing how contents of cells change place. (d) Section through skin of Hercules beetle, showing spongy layer containing either air or water.*

Flounders and other flat fish can adjust their pattern as well as their colour, to imitate the sea bed on which they lie, changing from a grainy effect, like Donegal tweed, when on sand, to a loud houndstooth check when against pebbles.

Rapid colour changes do not occur solely in response to the colour of the environment. Humans may blush when emotionally disturbed; so, too, does a male stickleback when sexually excited, and his eyes become brighter blue. When frightened, a chameleon will blacken, while an octopus, and many a human, will blanch.

The Hercules beetle changes colour in response to humidity, being black when damp, as at night, or when it is amid rotting bananas. In the dry day-time, it is yellow. The beetle's upper skin is a spongy, yellow layer which, when dry, scatters yellow light. Underneath it lies a black layer which shows through when the sponge gets damp, fills with liquid and becomes translucent. But this beetle, too, is emotionally reactive, and if it is frightened, it may change colour regardless of the humidity of its surroundings.

The mechanisms for all these colour changes are extremely complex and not yet fully understood. Those which result in camouflage colouring (and patterning) must be triggered by signals from the eyes or other light-sensitive organs. The way in which these signals are transmitted by the pigment cells varies. In squids and fish, the response is brought about largely by the nerves; in shrimps and frogs, largely by hormones; and in reptiles by a combination of methods.

Why, if we may use so difficult a word, are animals coloured the way they are? What good, if any, does it do them?

We can safely assume that, if a particular species of coloured animal has survived to the present day, its colour is no overriding disadvantage. Sometimes colour is immaterial. It matters little to vertebrates that their blood is red rather than colourless or blue, as in some lower animals; the crucial quality of a respiratory pigment is not its colour but its ability to supply the tissues with oxygen. The vivid veridian of its bones would seem to provide no advantage to the garfish, nor its allegedly rose-pink sweat to the hippopotamus. Many animal colours, however, seem to give the wearer a positive advantage which has contributed to the survival of the species.

Survival of an individual depends both on its obtaining its own food and on its not becoming food for someone else; and, for a hunting carnivore, inconspicuousness pays off. One is less likely to be eaten if one does not advertise oneself as a tasty morsel. This may involve advertising oneself as a non-tasty morsel, or being as inconspicuous as possible, or pretending to be a piece of abstract art. Some animals have evolved a realist approach, with a built-in assumption that the predator will get them anyway; so there is advantage in a 'start-eating-here' mark in some inessential part of the body. The occasional parasite, who *needs* to be eaten by its host, does indeed advertise itself as desirable food.

Inconspicuousness can be achieved in a number of different ways which have been exploited by much of the animal kingdom, including man (cf. page 205). Dappling is adopted by lion cubs and deer, who live among light undergrowth. Countershading is common in vertebrates and invertebrates alike; normally the top of the animal, which may be lit by the sun, is darker than the underside, which is in shadow. The resulting similarity in depth of colour of back and belly gives the animal a flat, somewhat unreal appearance. The caterpillars of some hawk-moths, however, have lighter backs than 'undersides'; since they walk *underneath* twigs, the same two-dimensional illusion occurs. Some animals have evolved a specialized resemblance to a particular background in both shape and colour. Some moths have wings extremely similar to the bark of one particular type of tree, and other insects are almost indistinguishable from curled leaves, bird droppings or dried sticks.

Mimicry of this type can evolve fairly quickly. Detailed work on the peppered moth has shown that, before the Industrial Revolution, the moth was usually a mottled 'pepper-and-salt' colour, well-camouflaged when settled on lichenous tree-trunks. In industrial areas, the lichen became replaced by bare, sooty trunks, and the moth was most commonly almost black, except in rural areas. Today, as the city air becomes cleaner, the paler form is making a modest comeback, even in regions with much heavy industry.

More flexible camouflage is achieved by those animals which can adapt their colour to their background (see page 96). Camouflage can also be achieved by breaking up the outline, so that the animal is not recognized for what it is. The broad, bright bands on many tropical fish may have this effect; the fish can easily be seen, but is not perceived as either predator or potential prey. To the human eye, they look more like abstract art than fish. The markings often contrast markedly with the background. Fish which inhabit the blue waters of coral reefs commonly have yellow markings, whereas fresh-water fish, who live in a green environment, are often splodged with red. The bands of wild piglets and many chicks, and narrower stripes of tigers and zebras disrupt the animal's outline in the same way. The interference and diffraction colours of beetles and bluebottles may also confuse predators, who might not expect their food to change colour like psychedelic lightning when it moves.

Iridescent colours may, on the other hand, serve to advertise their owners as untasty morsels. Many insects, particularly, are brightly coloured if they are of bitter taste or armed with a sting. Predators learn quickly, and one unfortunate encounter is often enough to protect the other members of the species from that individual hunter. Harmless insects often show a remarkable similarity to noxious ones, and no doubt enjoy considerable protection from the disguise. A predator who has once been stung by a wasp or bee would probably avoid a clear-wing hornet moth. Some animals may, when disturbed, produce warning colours which frighten off attackers. A chameleon may open its pink mouth, as well as turn black (see page 96). Puss-moth caterpillars raise their heads to reveal fearsome red markings, and there is a species of African grasshopper which intimidates such large aggressors as hens and monkeys merely by opening up its purple-black wings.

Some caterpillars are adorned with large eye-like patches to which predators react as if they had seen a snake. Eye-spots on the hind wings of moths and butterflies may also act as 'start-eating-here' labels. The head, thorax and abdomen, which between them contain all the vital organs, are usually drab. The blue tail of the blue-tailed skink fulfils the same role, and can, if eaten, be regrown. Colour in cock birds may serve to decoy a predator from the drabber female, sitting near by on eggs or chicks.

As well as determining an animal's appearance, its colour may also play an important role in the exchange of radiation between the animal and its surroundings. In humans, the melanin layer acts as protection from an excess of ultra-violet radiation. It is densest for races which originate in tropical parts, and in fair-skinned races thickens somewhat on prolonged exposure to sunlight. Some cold-blooded vertebrates, such as desert toads, are black when the sun is low and pale when it is high; when dark, they can absorb the heat they need, but at noon, they reflect the fiercest sunlight.

Colour also plays its part in communication between members of the same species. The bobbing white tails of retreating rabbits and deer serve as a danger warning to other individuals. Baboons make use of their coloured buttocks in a variety of social situations, including both sexual display and submission. Those species of bird in which the male is exotically coloured are often those with elaborate courtship displays, and it may well be that the coloured plumage is attractive to females. Herring-gull chicks peck at the red spots on their parents' bills, and this stimulates regurgitation into the open mouths of the young.

In humans, skin colour carries a wealth of social and cultural overtones and is possibly the most basic factor dividing (or uniting) man and man. Regrettable though this may be, it is not too difficult to see how it has arisen. It is rather easier to tell at a glance whether a man comes from Far Eastern stock than whether he has a deep love of higher mathematics. Nor is the labelling of persons by race any worse than the practice of labelling persons by any other criteria. It becomes undesirable only when unwarranted generalizations are made from the label, and deplorable when they form the basis of an inhumane act. Some generalizations are surely allowable; it seems that a very dense melanin layer is associated not only with frizzy hair and characteristic facial bones, but also with an enhanced sensitivity to musical beat. Maybe the time will come when we shall view these differences within our species as a source of enrichment rather than as a threat.

PART FOUR:
SENSATIONS OF COLOUR

The homogeneal Light and Rays which appear red, or rather make Objects appear so, I call Rubrifick or Red-making; those which make Objects appear yellow, green, blue, and violet, I call Yellow-making, Green-making, Blue-making, Violet-making, and so of the rest. And if at any time I speak of Light and Rays as coloured or endued with Colours, I would be understood to speak not philosophically and properly, but grossly, and accordingly to such Conceptions as vulgar People in seeing all these Experiments would be apt to frame. For the Rays to speak properly are not coloured. In them there is nothing else than a certain Power and Disposition to stir up a Sensation of this or that Colour. For as Sound in a Bell or musical String, or other sounding Body, is nothing but a trembling Motion, and in the Air nothing but that Motion propagated from the Object, and in the Sensorium 'tis a Sense of that Motion under the Form of Sound; so Colours in the Object are nothing but a Disposition to reflect this or that sort of Rays more copiously than the rest; in the Rays they are nothing but their Dispositions to propagate this or that Motion into the Sensorium, and in the Sensorium they are Sensations of those Motions under the Forms of Colours.

– NEWTON, *Opticks*

Since all Perception in the Brain is made
(Tho' where and how was never yet display'd)
And since so great a distance lies between
The Eye-ball, and the Seat of Sense within,
While in the Eye th'arrested Object stays
Tell, what th'Idea to the Brain Conveys?

– BLACKMORE, *Creation*

Do not the Rays of Light falling upon the bottom of the Eye excite Vibrations in the *Tunica Retina*? Which Vibrations, being propagated along the solid Fibres of the optick Nerves into the Brain, cause the Sense of seeing.

– NEWTON, *Opticks*

Two optic nerves, they say, she ties,
Like spectacles, across the eyes;
By which the spirits bring her word,
Whene'er the balls are fix'd or stirred.

– MATTHEW PRICE, *Alma*

Someone is given a certain yellow-green (or blue-green) and told to mix a less yellowish (or bluish) one – or to pick it out from a number of colour samples. A less yellowish green, however, is not a bluish one (and vice versa), and there is also such a task as choosing, or mixing a green that is neither yellowish nor bluish. I say 'or mixing' because a green does not become both bluish[1] and yellowish because it is produced by a kind of mixture of yellow and blue.

– WITTGENSTEIN, *Remarks on Colour*, trans. by McAlister and Schattle
[1] Translator's note: Wittgenstein wrote 'greenish' here but presumably meant 'bluish"....

During the day, owing to the yellowish hue of the snow, shadows tending to violet had already been observable; these might now be pronounced to be decidedly blue, as the illumined parts exhibited a yellow deepening to orange.

But as the sun at last was about to set, and its rays, greatly mitigated by the thicker vapours, began to diffuse a most beautiful red colour over the whole scene around me, the shadow colour changed to a green, in lightness to be compared to a sea-green, in beauty to the green of the emerald. The appearance became more and more vivid: one might have imagined oneself in a fairy world, for every object had clothed itself in the two vivid and so beautifully harmonizing colours, till at last, as the sun went down, the magnificent spectacle was lost in a grey twilight, and by degrees in a clear moon-and-starlight night.

– GOETHE, Theory of Colours

Colours appear what they are not, according to the ground which surrounds them.

– LEONARDO DA VINCI, Trattato della pi

In the silent painted park where I walked her and aired her a little, she sobbed and said I would soon, soon leave her as everybody had, and I sang her a wistful French ballad, and strung together some fugitive rhymes to amuse her:

> The place was called *Enchanted Hunters*. Query:
> What Indian dyes, Diana, did thy dell
> endorse to make of Picture Lake a very
> blood bath of trees before the blue hotel?

She said: 'Why blue when it is white, why blue for heaven's sake?' and started to cry again ...

– NABOKOV, Lolita

Why blue: when I asked Nabokov 'why blue?' and whether it had anything to do with the butterflies commonly known as the 'Blues', he replied: 'What Rita does not understand is that a white surface, the chalk of that hotel, does look blue in a wash of light and shade on a vivid fall day, amid red foliage. H. H. is merely paying a tribute to French impressionist painters. He notes an optical miracle as E. B. White does somewhere when referring to the divine combination of "red barn and blue snow". It is the shock of color, not an intellectual blueprint or the shadow of a hobby ... I was really born a landscape painter.'

– A. APPEL, The Annotated Lolita

Every hue throughout your work is altered by every touch that you add in other places.

– RUSKIN

When Anaxagoras says: Even the snow is black!
he is taken by the scientists very seriously
because he is enunciating a 'principle', a 'law'
that all things are mixed, and therefore the purest white snow
has in it an element of blackness.

That they call science, and reality.
I call it mental conceit and mystification
and nonsense, for pure snow is white to us
white and white and only white
with a lovely bloom of whiteness upon white
in which the soul delights and the senses
have an experience of bliss.

And life is for delight, and for bliss
and dread, and the dark, rolling ominousness of doom
then the bright dawning of delight again
from off the sheer white snow, or the poised moon.

And in the shadow of the sun the snow is blue, so blue-aloof
with a hint of the frozen bells of the scylla flower
but never the ghost of a glimpse of Anaxagoras' funeral black.

– D. H. LAWRENCE, 'Anaxagoras'

I had entered an inn towards evening, and, as a well-favoured girl, with a brilliantly fair complexion, black hair, and a scarlet bodice, came into the room, I looked attentively at her as she stood before me at some distance in half shadow. As she presently afterwards turned away, I saw on the white wall, which was now before me, a black face surrounded with a bright light, while the dress of the perfectly distinct figure appeared of a beautiful sea-green.

*

As the opposite colour is produced by a constant law in experiments with coloured objects on portions of the retina, so the same effect takes place when the whole retina is impressed with a single colour. We may convince ourselves of this by means of coloured glasses. If we look long through a blue pane of glass, everything will afterwards appear in sunshine to the naked eye, even if the sky is grey and the scene colourless. In like manner, in taking off green spectacles, we see all objects in a red light. Every decided colour does a certain violence to the eye, and forces the organ to opposition.

– GOETHE, *Theory of Colours*

A given visual phenomenon may not be perceived at all unless it is actively looked for.

– BURNHAM, HANES and BARTLESON, *Color*

The diff'rence is as great between
The optics seeing, as the objects seen.
All Manners take a tincture from our own;
Or come discolour'd through our Passions shown.
Or Fancy's beam enlarges, multiplies
Contracts, inverts and gives ten thousand dyes.

— POPE, *Moral Essays*

For they sometimes appear by other Causes, as when by the power of Phantasy we see Colours in a Dream, or a Mad-man sees things before him which are not there; or when we see Fire by striking the Eye, or see Colours like the Eye of a Peacock's Feather, by pressing our Eyes in either corner whilst we look the other way. Where these and such like Causes interpose not, the Colour always answers to the sort or sorts of the Rays whereof the Light consists, as I have constantly found in whatever Phaenomena of Colours I have hitherto been able to examine. I shall in the following Propositions give instances of this in the Phaenomena of chiefest note.

— NEWTON, *Opticks*

11

LIGHT AND THE EYE

The colours we see depend, we say, on the composition of the light which enters the eye. And this, in turn, depends on the composition of the original light and on the way in which the light is modified by encounters *en route* from source to eye: encounters with other light waves, with the media through which it passes and with objects off which some or all of it may bounce. Since we were less than two weeks old, we could experience variations in the composition of the light as different sensations of colour. But how?

In this section, we shall discuss the changes which occur when the light encounters the back of the eye. We shall discuss mainly human colour vision, both normal and anomalous, but will also mention colour vision in other vertebrates and in insects. We shall see that there is still much unravelling to be done before we can understand what happens *after* the light is absorbed by the eye: the optic nerve and the brain combine to play some odd tricks. Some of the sensations we experience seem but tenuously related to the encounter between light and eye. So, while it is quite easy to give the specifications (of wavelength and intensity) for a ray of light, it is vastly more difficult to specify a sensation of colour. None the less, as we shall see, there have been numerous brave attempts to do so.

The eye is commonly likened to a camera. The light enters a dark chamber through an aperture which can be varied in diameter according to the intensity of the light, and it is focused by the lens on to a light-sensitive backing. In both camera and eye, the light produces only small changes in the photosensitive material, but these are the starting points for a long series of changes which take place during the 'processing' and which eventually produce either a photograph or a sensation.

In a young, normal, human eye (see Figure 43), the outer layer (the 'cornea'), the lens and the eye fluids are almost transparent. When the light reaches the retina, on the inner surface of the eye, it passes through layers of transparent nerve fibres, between capillary blood vessels and on to the photosensitive cells at the ends of the nerve fibres. These are backed with a layer of cells which contain black pigment and absorb stray light. The whole eye is enclosed in a tough outer skin which is an opaque continuation of the cornea.

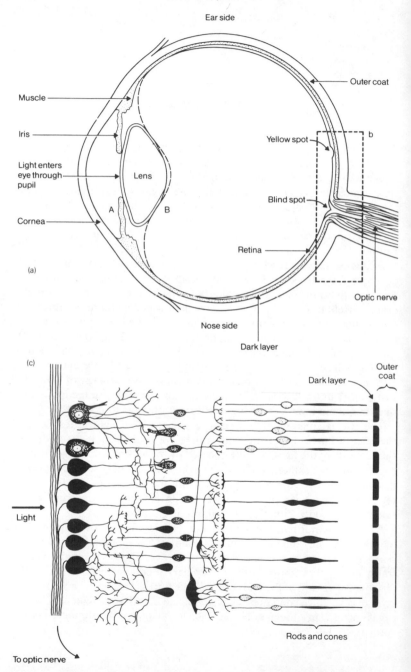

(a)

Ear side

Outer coat

Muscle

Iris

Light enters eye through pupil

Lens

Cornea

A

B

Yellow spot

Blind spot

b

Retina

Optic nerve

Nose side

Dark layer

(c)

Outer coat

Dark layer

Light

To optic nerve

Rods and cones

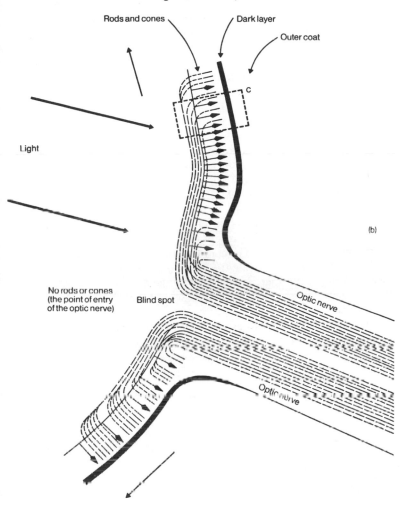

Figure 43. The human eye *(a) Horizontal section through centre. The cavities A and B contain fluid. (b) Enlargement of region shown in box in (a) above. (c) Enlargement of region shown in (b) above, showing section through retina. (Adapted, with permission, from Figures 2.17, 2.18 and 2.19 in R. W. Burnham, R. M. Hanes and C. J. Bartleson, Color, John Wiley, New York, 1963.)*

The photosensitive cells are coloured; they contain pigments which absorb visible light; and it is this absorption which forms the basis of our sense of sight. In the human retina, there are two classes of photosensitive cells called *rods* and *cones* on account of their (very approximate) shape. The rods are effective only in dim light and enable us to sense differences in brightness, while the cones respond to light of normal intensity and allow us to distinguish between different colours. So the eye behaves as a camera which contains two films, one for colour and one for black and white only, as well as a photochemical device which selects the film appropriate to the lighting conditions.

The rods and cones are not distributed evenly over the retina. Opposite the centre of the lens is a small yellow-brown pit, the centre of which (the 'fovea') contains only cones; 100,000 of them. There are no capillaries or nerve fibres between the lens and these foveal cells, but the pigmented edge of the pit absorbs some of the ultra-violet and blue light which might otherwise damage them. The rest of the retina is equipped with 4 million cones and 120 million rods. The concentration of the rods is greatest about 20° from the yellow spot, while that of the cones decreases with increasing distance from it. But there is one point on the retina, at the junction with the optic nerve, where there are no photosensitive cells of either sort. This is the 'blind spot'.

When light strikes a photosensitive cell, a photon may be absorbed, and if it is, it will trigger off a series of changes which contribute to the sensation of

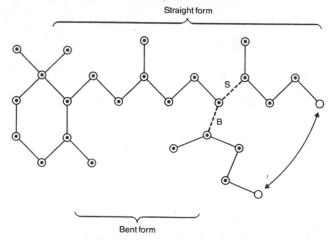

Figure 44. A light twist *The skeleton of the light-sensitive substance retinal (cf. carotene, shown in Figure 36, page 87). The end of the tail can twist, to give a bent form joined to the backbone by the dotted line B, or a straight form, attached by line S.*

vision. The probability of the photon being absorbed depends on whether the eye is adapted for bright or dark conditions, on the wavelength and intensity of the light and on the type of retinal cell on which it falls. We shall first see the way in which the rod cells respond, since these operate more simply than the cones.

It seems that all rods contain a single reddish-purple pigment called 'visual purple' or *rhodopsin*, which consists of a protein combined with a substance called retinal, which, like carotene, has a long backbone of carbon atoms, with small side-arms (see Figure 44). When we have been in

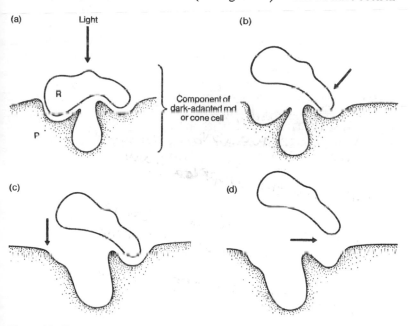

Figure 45. Changes in the retina *Sketches to illustrate the mechanism of light sensitivity in the eye.*

(a) A dark-adapted retinal group R adopts the bent form and fits snugly on to the puckered surface of a protein, P.

(b) When the retinal absorbs a photon, its 'tail' straightens and forces the 'head' away from the protein base.

(c) The protein surface, no longer held in position by the retinal, relaxes, and changes shape.

(d) The 'tail' of the retinal breaks away from the protein, initiating an electrical charge which is conveyed to the nerve.

(Adapted, with permission, from R. Hubbard and A. Kroft, Scientific American, *vol. 216, June 1967, p. 65.)*

the dark for a while, the carbon chain is bent and slightly twisted near one end, and able to fit into an indentation in the surface of the protein (see Figure 45). If dim light enters the eye, a photon may be absorbed by the carbon chain, which then straightens out, untwists and detaches itself from one of its moorings on the protein. This causes the shape of the protein indentation to change, and the other end of the chain to break loose. The pigment is bleached, and a signal may pass along the nerve fibre. This signal is identical for every proton absorbed, regardless of the wavelength of the light. The rod response is like that of a spring mousetrap which behaves exactly the same whether the trigger is released by a mouse or a rat (although the probability that it will be set off by one or other depends on the relative numbers of each around and their relative liking for the bait). The rod cells are not, in fact, equally sensitive to light of all wavelengths; photons in the middle of the visual region are absorbed most efficiently (see Figure 46). When viewed by weak moonlight, the dark green leaves of a holly tree might seem a lighter grey than the bright red berries, because rods show their greatest sensitivity to green light.

If the illumination is increased, the rhodopsin remains bleached and the rod cell is inactive. But, in dim light, the pigment, having lost its energy to

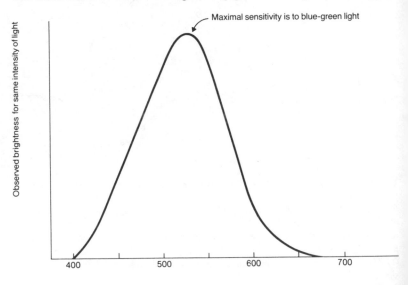

Figure 46. Brightness by night *How the rod cells respond to light of the same intensity but of different wavelength.*

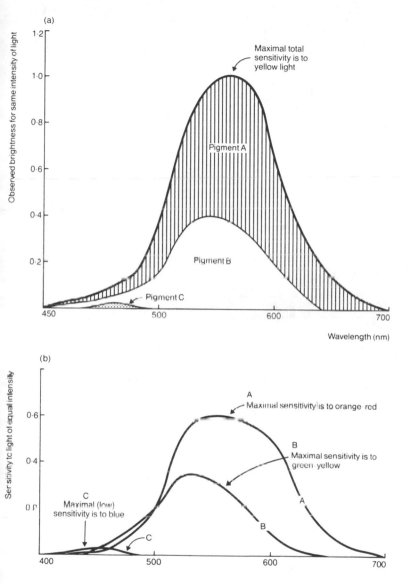

Figure 47. Brightness by day *(a) The full line shows the combined response of the cone cells to light of the same intensity but of different wavelengths. The approximate contributions of the three systems of cone pigments are indicated by the differently shaded areas. (b) The approximate sensitivities of each of the separate systems.*
The valves are those estimated at the cornea by Smith and Pakorny.

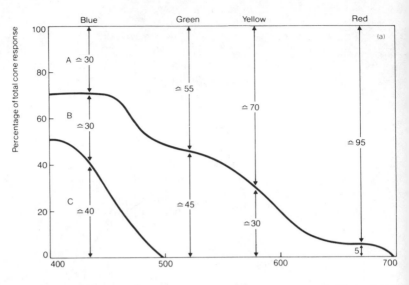

(a)

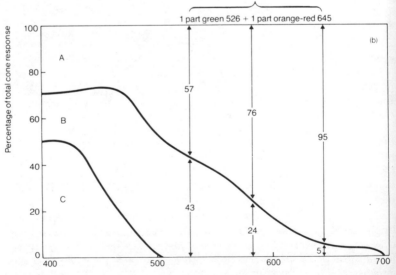

(b)

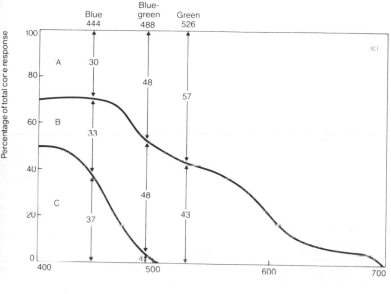

Figure 48. Cones and colour: *Approximate △ percentage of total cone response attributable to each of the three pigments.*

(a) *The probable cone response ratios A:B:C which provide the sensations of blue, green, yellow and red.*

(b) *The sensation of yellow can also be obtained by combining the responses of cone pigments A and B to green light and to orange-red light.*

(c) *The blue-green sensation obtained from light of 488 nm can never be exactly matched by mixing blue and green light as the percentage of B-cone response is maximal in this region.*

the nerve, gradually reverts to its original, photosensitive state. The mousetrap resets itself spontaneously. Rods which have been inactivated by bright light readapt only slowly to the dark. The efficiency of our night vision increases markedly over twenty or thirty minutes and is not fully developed for about one hour.

Illumination which is bright enough to inactivate the rods is also bright enough to stimulate the cone cells and so to allow us to perceive colour. Cones respond to changes in the overall intensity of the light much more quickly than do rods and are fully adapted within seven minutes.

Sensations of Colour

Each cone, like each rod, contains visual pigment, a molecule of which consists of a carotene-like carbon 'backbone' partially folded into a cavity in a protein molecule. Although the retinal backbone seems to be exactly the same as that in a rod pigment, the particular protein which forms part of rhodopsin has not been found in cones. In fact, there are three different types of cone, each containing a different sort of protein. So a cone will have one of three possible visual pigments. Each pigment absorbs light over much of the visual spectrum (see Figure 47a). The effect of the absorption of a photon is similar to that in rhodopsin, and does not vary with the wavelength of the light: for each photon absorbed, one carbon backbone straightens and one impulse passes to the nerve.

Each cone pigment, like rhodopsin, absorbs photons with varying efficiency according to their energy. Colour vision is possible because the three cone pigments differ *from each other* in their sensitivity to wavelength.

We can see from Figure 47b that one of the cone pigments (A) absorbs most efficiently in the orange part of the spectrum, while another (B) absorbs maximally in the green region. The third pigment (C), which absorbs less efficiently than the others, has its absorption peak in the blue, high-energy end of the spectrum. When we add together the separate contributions of the three pigments, we find (Figure 47a) that the total absorption also varies with wavelength, being most efficient in the yellow region. The percentage contribution of each pigment to the total absorption is shown in Figure 48, which perhaps provides us with the best key to the understanding of colour vision, because it shows us how the ratio of contributions varies with wavelength. Any light of 'normal' intensity triggers off the three cone pigments in some ratio A:B:C which depends on the wavelength composition of the light: and it is this *ratio* which produces a particular sensation of colour. For example, a cone response in a ratio 33:42:25 gives a sensation of blue, while with one of 70:30:0 we perceive yellow (see Table 2).

Some response ratios can be produced both by the light of a single wavelength, and by a mixture of two lights, both of different wavelengths. We cannot distinguish by eye alone between a 'pure' yellow light of one narrow wavelength range in the region of 580 nm and the mixture of green and orange-red lights which gives the same response ratio (see Table 3 and Figure 48b). Green light of 560 nm cannot, however, be matched by a mixture; since this wavelength is the lowest at which pigment C makes no appreciable response, any mixture containing light of lower wavelength would stimulate some contribution from pigment C. Nor can blue-green light of 488 nm, where the response of B-cones is maximal (see Figure 48c).

There are also some 'non-spectral' colours which can be made only by mixing. Mixtures of red and blue light give response ratios which do not

Table 2. *Probable response of cone pigments to various wavelengths*

Wavelengths (nm)	Approximate % of total absorption contributed by pigment			Colour
	A	B	C	
420	29	21·5	49·5	violet
460	33	42	25	blue
490	48	48·5	3·5	blue-green
530	57	43	0	green
580	70·5	29·5	0	yellow
600	80	20	0	orange
620	88·5	11·5	0	orange-red
660	95·5	4·5	0	red

Table 3. *Probable cone response ratios for some pure and mixed lights*

Colour	A:B:C ratio	Wavelength of pure light of same ratio (nm)	Mixed light of same colour			
			%	nm	ratio	colour
Yellow	70:30:0	588	50	526	57:43:0	green
			50	645	95:5:0	orange-red
Blue	33:42:25	460		None		
Blue-green	48:48:4	488		None		
Green	63·5:36·5:0	560		None		
Red	95·5:4·5:0	660		None		
Magenta	64·5:23:12·5	None	50	460	33:42:25	blue
			50	660	95·5:4·5:0	red
White	63·5:34:2·5	None	34	480	43:49:8	blue
			33	540	59:41:0	green
			33	620	88·5:11·5:0	orange-red

correspond to that of any single wavelength light and which we perceive as shades of magenta (see Table 3). And if we mix blue, green and orange-red light of roughly equal intensities, we obtain white light of ratio about 63·5:34:2·5, which is almost indistinguishable from the white light obtained by mixing all visible wavelengths at equal intensity.

So the cones allow us to distinguish colours by day and the rods enable us to see shapes by night. But the rods sometimes also contribute to vision by day. An object at the very edge of our visual field (behind the lines joining one eye to opposite ear) seems monochrome because the light which reaches us from it falls on the periphery of the retina where there are mainly rods and only few cones (see Figure 52). Slightly nearer the centre of the retina, there are rather few cone cells, and the perceived colour is not quite the same as that produced when the same light falls on the cone-rich central region; it has been suggested that the sensation is attributable to the combined response of rhodopsin and of the three cone pigments.

The responses of rods and cones certainly both contribute to vision at the intermediate, twilight, intensities of illumination. When the light is bright enough for us to see some colour, but not bright enough to bleach the rods totally, the sensation is probably a result of the response of all four pigments. We have seen that rods absorb most efficiently in the green region of the spectrum (~ 500 nm), whereas the total cone response is most sensitive to yellow light (550 nm). As dusk falls, our eyes therefore become increasingly sensitive to light of shorter wavelengths. At twilight, our most sensitive response is to blue-green light (510 nm), and so, before all colours turn to greys, they become gradually more blue, an effect first reported in 1823 by the Czech physiologist Purkinje.

Night is not always devoid of colour. But although the full moon looks golden,* the moonlit world is colourless, as Walter de la Mare expresses it:

> Slowly, silently, now the moon
> Walks the night in her silver shoon
> This way and that, she peers and sees
> Silver fruit upon silver trees.

The intensity of sunlight reflected from the moon is just high enough to stimulate cone vision, at least in that part of the spectrum to which cones are

---

* But the moon is not always golden. Occasionally an unusually high proportion of the blue component of the reflected light is scattered, or bent, away from the observer. A 'harvest moon' or 'hunter's moon' (or a rising sun) low in the sky looks redder than usual. During a lunar eclipse the colour of the moon may change, through orange to deep crimson. And, once in a blue moon, atmospheric particles from a volcanic eruption or forest fire may be of exactly the right size to scatter all light of low and medium energy, with the result that only the blue component reaches the observer.

most sensitive. But the intensity of the light which is reflected off other objects is so low that only rod vision occurs.

The stars, too, look silver, although spectroscopic analysis shows that they radiate light of composition which is often far from white, although too dim for our eyes to recognize it as coloured. The colours of the stars have, however, been photographed using very long exposures.

Both rods and cones have low sensitivity to light of long wavelength, and it is likely that the red light of a photographic dark-room, though strong enough to stimulate the cones, does not inhibit the rods. Sailors and airmen who need their eyes to become adapted for dark vision before going on night duty can grow acclimatized by wearing red goggles in a lighted room instead of having to stay in the dark. The US armed forces are issued with playing cards specially designed so that hearts and diamonds, normally represented in red on a white ground, may be distinguished by those wearing such goggles.

The low sensitivity of the retina to long wavelengths leads to the curious behaviour of certain green celluloid eyeshades. Although, seen through a single layer of celluloid, the world looks green, through a double layer, it looks red; and similar effects can be obtained with certain coloured liquids. When white light falls on the eyeshade, all the very low-wavelength light is transmitted together with about one third of the green light, and rather little of other wavelengths. Since the retina is so much more sensitive to green light than to red, the resultant sensation is green, albeit a somewhat yellowish green. If this 3:1 mixture of red to green light passes through a second layer of celluloid, the red component is again transmitted almost undiminished, while the strength of the green is again reduced threefold. The 9:1 ratio of red to green in the emergent light is high enough to produce a red sensation, despite our low sensitivity to long wavelengths.

12

ANOMALOUS COLOUR VISION

The normal human eye is by no means the only one capable of discriminating between light of different wavelengths. Many humans who have abnormal colour vision are none the less able to differentiate between light of different wavelengths, as are some species of animals.

There is considerable variation even in 'normal' human vision. Some is individual, and some racial, as in the pigmentation of the lens and of the yellow spot. Changes may encroach with age, as when the cornea becomes opaque and perceived colours dull. Those successfully recovered from an operation for cataract often report a dramatic brightening of perceived colour. The lens pigment darkens with age, and so absorbs a higher proportion of short-wavelength light. Many artists use blue less frequently as they get older; and Mark Twain compared some of Turner's later work to 'a ginger cat having a fit in a bowl of tomatoes'. Defective colour vision in humans is occasionally caused by injury, drugs or physical or mental illness. But of the 8 per cent of all men and 0.5 per cent of all women who have abnormal colour vision, the great majority are born that way.

Albinos, who lack pigment in their skin and hair, also lack the black layer behind the retinal cells, which are therefore stimulated by stray light scattered inside the eye as well as by light from outside. Their vision therefore lacks clarity and the colours become diluted and less brilliant. The occasional person who is born without any cone cells will have very poor general sight. Relying on rod vision, he will shun bright light and be totally colour-blind. But most types of defective colour vision are the consequence of abnormalities in the cone cells.

Human colour vision has been widely investigated by means of experiments in which an observer is asked to match a given patch of light by mixing, in any proportion, a number of other lights. An observer with normal vision can match *any* colour, provided he has up to three others and can, if he wishes, mix one with the sample patch and then match this with a mixture of the other two lights. The need for three independent lights is a consequence of the presence of three cone pigments.

Some of the observers with abnormal vision can match any given colour by using only two lights, suggesting that they have only two photosensitive cone pigments. A very few people match by brightness alone, needing only one light; and about 6% of the male population need three, but produce matches which are quite unacceptable to those with normal vision. So it seems that defensive colour vision can arise either because one, or even two cone pigments are absent, or because all or any of the cone pigments are different from the normal types.

The most familiar forms of defective colour vision arise from the absence of one of the three cone pigments. To about 2 per cent of all men, green and red are both indistinguishable from grey. Others, who can discriminate between red and green, confuse yellow, blue and grey; and yet another combination of two cone pigments allows normal discrimination of red, orange, yellow and green, although all light of shorter wavelengths appears blue-green.

We can see from Figure 48 how colour vision would be affected if we

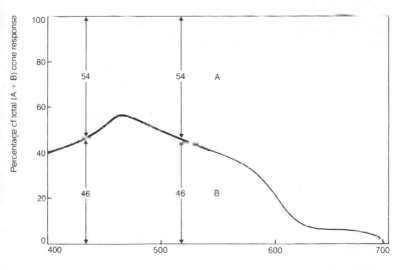

Figure 49. Blue-green ambiguity *The likely effect of absence of pigment C on the total cone response, if the sensitiveness of pigments A and B were as represented in Figure 47 (page 115). For light of wavelengths above about 550 nm (thin curve), colour vision would be unaffected. But at lower wavelengths (thick line), the same response can be obtained from pure lights of two wavelengths (e.g. 430 nm and 573 nm) and of any mixture of them.*

lacked one of the three cone pigments while the other two were unchanged. If either A or B was missing, colour discrimination would be possible at wavelengths below about 520 nm, where two pigments respond. At higher wavelengths, however, only one pigment is active, and only differences in brightness could be perceived. On the other hand, if C were missing, there would be response from two pigments over the whole visual range. Colour discrimination would be possible except in those regions where the vertical distance between the curves happened to be the same. Figure 49 indicates that cones containing only pigments A and B might give an ambiguous response over the wavelength range 420–520 nm (blue-green).

It is often emphasized that the colour vision of those who are classified as 'colour-blind' is often only slightly more defective than that of the 'normal' majority, who, as we have seen (page 118), cannot distinguish between, for example, 'pure' yellow light and a mixture of red and green lights; or, indeed, between a large number of other pairs of 'pure' and 'mixed' light. Those with defective colour vision merely have more scope for ambiguity than do those with the normal, though far from perfect, ability to discriminate between light of different wavelengths. Since some jobs, however, involve the recognition or matching of colours, tests for anomalous colour vision have been devised. It is only from tests that many of those with defective colour vision learn that they have any abnormality. There are not a large number of situations, other than work, in which a tendency to muddle green and blue would be apparent, particularly as many of the normally sighted differ about whether a given hue in this range should be called blue, turquoise or green (see Chapter 21). The tests may be strictly tied to the job and designed only to find out whether an applicant can match paint, cloth, printing ink, recognize signals or select the correct, colour-coded electrical resistor. Other more general tests may involve the naming of coloured lights and sorting large numbers of coloured samples into regular series involving a gradual change in hue. One common type of test uses cards covered with an irregular mottled display of spots in which are embedded others depicting numerals, letters, pathways or simple pictures. The figures differ from the background in colour but not in lightness and are designed so that each can be read only by those who can distinguish between the two colours of a pair. A set of such cards enables most types of defective colour vision to be detected. If the equipment is available, a person may be asked to match one light with a mixture of three, or two, others. Information about this ability to distinguish colours is given by the match which he selects, particularly if he is asked to match the same light a number of different times.

Most defective colour vision is hereditary. Red-green colour 'blindness' is transmitted by a recessive gene, sex-linked to the male. So a male who has one gene carrying this defect will, in fact, be colour-blind, while a female will not be colour-blind unless she has two genes for colour-blindness; and such genetic 'purity' is rare. But a woman with one such gene may transmit the defective gene to her children, although she herself will have normal vision. So her sons may be colour-blind and her daughters may, like their mother, be carriers. Men, if they have normal colour vision, cannot transmit the defect; a colour-blind son of a normal father must have inherited the defective gene from his mother.

It is not surprising that the incidence of a genetic defect of this type shows a regional variation. In some rather isolated communities, such as Fiji, it is very low (only 0·8 per cent of the male population), whereas in Canada it is as high as 11·2 per cent.

With nearly two centuries of hindsight, we can appreciate the observation reported by John Dalton, the chemist, that the geranium described by his friends as pink appeared, both to him and his brother, as sky blue by day and 'red' by candlelight. He must have lacked the red sensitive pigment, so that the mixture of red-blue light scattered from the flowers would be indistinguishable from the blue of the sky; and, in the candlelight, which contains little blue, the flowers, like any red object, would scatter little light that he could detect and so would look black. Anomalous colour vision of this type was long termed 'Daltonism'.

13

COLOUR VISION IN ANIMALS

It is, of course, much easier to study colour vision in man than in other animals. But we can try to find out if animals can distinguish between light of different wavelengths by attempting to train them to seek food from a container of a particular colour, regardless of the brightness of the colour or the position of the container. Experiments can also show whether or not a change in the wavelength of the light produces a different electrical response from the nerve of a light-sensitive cell, although the results of such measurements alone cannot tell us whether the animal experiences different colour sensations, as these depend on the connections between the nerve cells as well as on the sensitivity of the eye to wavelength.

The great majority of studies of colour vision have been on either vertebrates or insects. Although most vertebrates have simple eyes, very similar in structure to the human eye, they vary enormously in their ability to distinguish between colours. Mammals other than man often have poor or non-existent colour vision, although that of the macaque monkey is very similar to our own. Ground squirrels, who can distinguish between blue and green but are insensitive to red, are thought to have only two types of cone cell. The red squirrel and the guinea pig have only one type of retinal cell and so are totally colour-blind. Cats have very poor colour vision, but, under good conditions, can distinguish blue-to-green colours from orange-to-red ones. It is thought that the proverbial red rag infuriates the bull by its movement rather than its colour, because no cones are present in the eyes of cattle.

Many types of bird and fish are thought to have good colour vision, since colour plays an important part in some of their behaviour. The Australian bower bird decorates its nest with various blue objects such as scraps of paper and china, juice from blue berries and feathers from smaller birds it has killed for the purpose. A male robin will defend its territory against a shapeless bunch of red feathers, although not against a bunch of brown feathers or even against a stuffed juvenile (brown) robin. Owls, however, have only rod cells, and so are totally colour-blind. Three types of retinal cone cells have been found in hens and pigeons, whose colour vision is similar to our own. The eyes of many birds, including these two, contain coloured 'filters' of oil droplets in the yellow-to-red range, but the effect of these on the colour vision of birds is not known. In the frog, colour vision appears to come with maturity; tadpoles seem to be colour-blind. Lizards, which are active only in daylight and have pure cone vision, seem able to distinguish between meal worms dyed different colours, while the nocturnal gecko, with pure rod vision, is colour-blind. Turtles, like birds, have coloured, oil-drop filters, but their colour vision seems to differ greatly from one species to another. Fish seem well able to distinguish colours. Sticklebacks and Siamese fighting fish react vigorously to red and blue in both courtship and defence of territory.

Though the compound eyes of insects have a totally different structure from the simple eyes of vertebrates, they contain visual pigments similar to our own. Many insects, such as bees, wasps, ants, dragonflies, butterflies, moths, beetles, cockroaches and houseflies, can distinguish colours, but usually in a range different from our own. And some have colour vision only in certain directions but are colour-blind in others. Some parts of their compound eyes are sensitive to wavelength and other parts are not. Many colour-sensitive insects respond to radiation of wavelength 390–300 nm in the ultra-violet range which is invisible to humans. But most insects are insensitive to red light. Exceptions are a few butterflies and the firefly, which can detect light of a wavelength as long as 690 nm; as indeed it would need to if, as believed, it uses the orange-red flashes of its species to find a mate. Ants, however, are blind to red. When a colony is illuminated by a spectrum from sunlight, they carry their larvae into the red region but avoid the near ultra-violet. In addition to colour vision, many insects, unlike humans, can distinguish between light which is of the same wavelength but polarized in different planes.

Much work has been done on the colour vision of bees, who can distinguish between lights in the three colour ranges: yellow (590–500nm), blue-green (500–480nm), ultra-violet (400–300nm) (see Figure 50). For bees, the mixture of 440 and 360nm is indistinguishable from violet (400nm). Bees also recognize as a separate colour the mixture of their two extreme visible wavelengths, 300nm and 650nm, known as 'bees' purple' by analogy with the mixture of the two extreme colours, red and violet, visible to humans. For bees, as for humans, there are pairs of complementary 'colours' which together make up the whole range of the visible part of sunlight: ultra-violet and blue-green, yellow and violet, blue and 'bees' purple'. Since bees cannot see light of a wavelength longer than about 590nm, they are blind to red and orange, and so it might seem surprising that they visit such red flowers as poppies. However, poppies reflect not only red light but also ultra-violet, and it is this which attracts the bees.

Photographs on film which is sensitive to ultra-violet light reveal complex

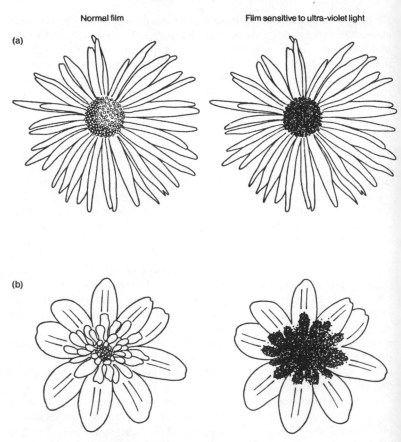

Normal film   Film sensitive to ultra-violet light

(a)

(b)

Figure 50. A bee's-eye view? *The photographs on the left were taken with normal film which responds mainly to visible light; those on the right were taken with film which records predominantly ultra-violet radiation. The outer petals of the flowers, which seem yellow to us, reflect 'bee's purple' to the bee. The centres, which absorb ultra-violet and reflect only the yellow, would seem much darker, and could guide the bee to the nectar.*
(a) Leopard's bane.
(b) Lesser celandine.
(Reproduced with permission, from R. L. Gregory and E. H. Gombrich, Illusion in Nature and Art, *Duckworth, London, 1973.)*

patterns of guide-lines on petals of some flowers which to us look plain white or a single colour, but which are 'variegated' in their power to reflect ultra-violet radiation. Similar 'latent' patterns on butterfly wings are thought to play a part in recognition and in courtship. If bees are shown two blackboards bearing the directions 'Bees may feed here' and 'No bees allowed', the bees will always collect on the correct notice, provided that the permissive text is painted with a material which reflects ultra-violet light while the restrictive one is painted with chinese white, which does not.

Although we have been able to establish that some animals can distinguish colours, only a few species have been studied in detail; and the part which colour plays in their lives is largely a matter of speculation.

14

THE EYE AND THE BRAIN

Are the writers of the excerpts on pages 104–7 merely being fanciful? If not, how can we reconcile the phenomena they describe with our earlier interpretation of colour vision in terms of the absorption of photons by the three cone pigments? Our own experience confirms that there is more to colour than the composition of the light which meets the eye. Colours provoke emotional responses; they appear to vary when physics suggests they should not and, contrariwise, seem constant when it seems they should vary; we sometimes see colours in the absence of what we might think would be the appropriate light, and, indeed, in the absence of any light at all. A great number of types of change can take place both in the retina and in the brain after a photon is absorbed.

Some of our psychological responses to colour may have a simple, geometrical origin. A red splodge often seems to advance from the page and to concentrate the viewer's attention towards its centre, whereas a blue splodge seems to recede and to lead the eye outwards. The apparent movement of colours towards, and away from, the viewer probably occurs because the lens of the eye (unlike that of a camera) is made of a single material and so is subject to 'chromatic aberration'; light of different wavelength is bent, as by a prism or a raindrop, to different extent. When the muscles which control the shape of the lens focus green light on the retina, the red light is focused a little behind it and the blue a little in front (see Figure 51). To see a red object clearly, the lens must be the same shape as we need to see a green object which is slightly nearer; and to see a blue object, it must be adjusted in the opposite way. So reds seem to advance and blues to recede.

The impression that blues (and yellows) spread while reds contract may well be because of the way in which the cone pigments are distributed on the retina. We cannot see colours at the edge of our field of vision because only rods are present at the periphery of the retina (see page 120), but as we bring an object nearer to the centre of the visual field, we first recognize blue and yellow. Slightly nearer the centre, we sense green, and still more centrally, we can see red (see Figure 52). It is not surprising that blue, which we can see at the widest angle of colour vision, tends to spread, and that red, which is visible only fairly centrally, tends to contract.

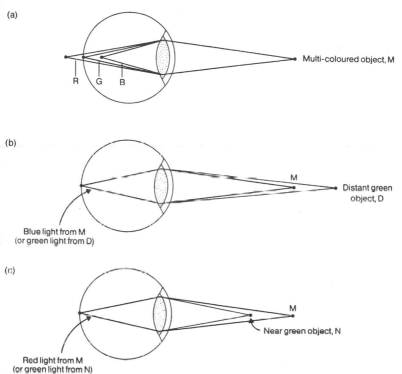

Figure 51. Colours which recede or advance *(a) The eye receives light from a multi-coloured object, M, when the lens is adjusted to focus green light on the retina. Blue is focused slightly in front of it and red slightly behind it.*

(b) When the lens changes to focus the blue light from M on to the retina, the eye is adjusted for the clear viewing of green (or white) objects, D, which are slightly more distant than M. Blue therefore seems to recede.

(c) Adjustment of the lens to focus red on the retina also allows clear vision of a green or white object, N, which is slightly nearer than the source of the red light. So red seems to advance towards the viewer.

Colour sensations may depend on intensity as well as on wavelength, even when the cones alone are responding. As the light gets more intense, both orange and yellow-green approach yellow, while violet and blue-green both become bluer. Only in three cases, yellow, green and blue, does the colour seem to be independent of the intensity. These colours are termed 'psychological primaries' because each can be said, whatever its intensity, to contain no element of any of the others.

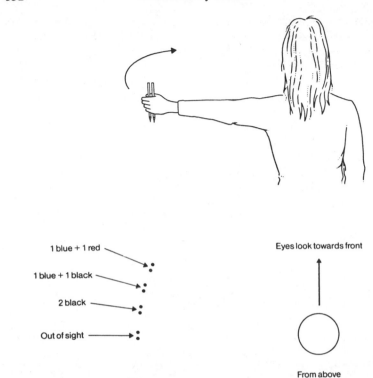

Figure 52. The corner of the eye *The girl, facing forward, holds a blue pencil and a red pencil out beyond her ear, and gradually moves her arm so that they come into her field of vision. She first sees both pencils as black, then perceives the colour of the blue one, and finally recognizes the red.*

We do not fully understand why, alone of all the colours, red, yellow, green and blue behave in this way, though the cause probably lies in the absorption curves of the cone pigments. Indeed, there are many large gaps in our knowledge of visual perception, and particularly of colour. But many of the effects mentioned in this chapter arise either from the highly complex set of nerve connections between the retina and the brain; or from the time lag in the recovery of a cone cell after it has absorbed a photon; or from a combination of both factors.

The rods and the cones are incorporated into very different 'circuit diagrams'. Impulses from individual rod cells do not travel directly to the brain, but first pass from a group of cells to a nerve centre which transmits a signal to other centres, from which an impulse eventually reaches the brain. Since the signal may have come from any one of a hundred or so cells, rod vision is very sensitive, although its definition is not of the very highest. It is likely that impulses from adjacent nerve centres are combined to give information about the *differences* between their signals, which enables them to monitor contrast, and us to recognize shapes (see Figure 53).

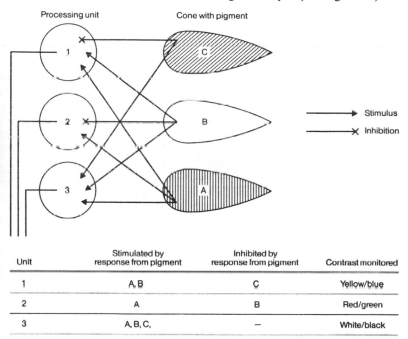

Unit	Stimulated by response from pigment	Inhibited by response from pigment	Contrast monitored
1	A, B	C	Yellow/blue
2	A	B	Red/green
3	A, B, C,	—	White/black

Figure 53. Connections for contrast *A possible scheme of nerve connections which would allow cone cells to monitor contrast.*

Signals from the cones travel to the brain by various routes. Some cone cells near the centre of the retina send combined responses straight to the brain, and this allows us to see extremely clearly, in colour, in light of normal intensity. Other cones seem to combine their signals so that the activity in one cell inhibits the response of neighbouring ones. The cones seem to be able to monitor both brightness and colour (red/green and blue/yellow), and this could be achieved if the nerve connections were similar to those of the simplified scheme shown in Figure 53. A strong response from the cones must also be conveyed to the rod cells, in order to inhibit them when the light is bright.

It is thought that the first steps in the collation of cone responses occur in the retina itself, but that the brain performs most of the processing required to produce the sensation of colour. It is certainly the brain which combines the signals from each eye. Coloured stereoscopic pictures provide impressive application of binocular vision. A scene is photographed twice; once as viewed by the right eye of a stationary observer, and once as viewed by his left eye (see Figure 54). The right-hand one, printed in red, is superimposed on the left-hand one, printed in blue-green. This composite print is viewed through goggles with a blue-green filter over the right eye

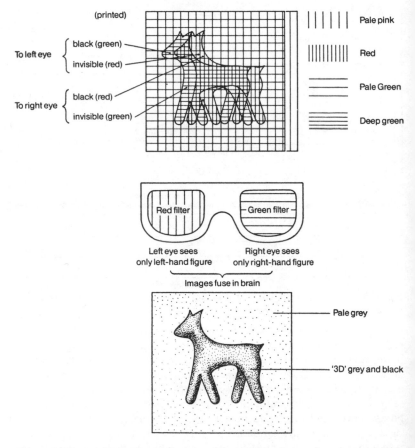

Figure 54. Stereo *An illusion of depth is produced by the fusion of two images within the brain.*

and a red filter over the left one. So the right eye sees black on blue-green, while the left eye sees black on red. After a second or so, the two signals fuse in the brain to provide a gleamingly three-dimensional impression in black on white.

Some people find that impressions from their two eyes do not fuse easily, if at all. The response of one eye sometimes entirely dominates that of the other, so that, when viewing a stereoscopic picture through coloured goggles, they would see a flat picture on a coloured background. A very few people, who have defective colour vision in one eye only, are in the unusual position of being able to compare the sensations produced by normal, and defective, colour vision, and to tell the normally sighted how the world looks to the 'colour-blind'. And if the detective eye is strongly dominant, their supposedly binocular colour vision will also be defective.

Some of the colour phenomena which have most captured the attention of writers and experimentalists alike are those in which the effect of light on one cone cell is modified either by the simultaneous response of neighbouring cells, or by some previous response, in eye or brain. One such effect is the way in which the inhibition of one cell by another enhances colour contrast. If we stare at a bright red patch on a white card, it seems to develop a green line around the edge, whereas a yellow border appears around a violet patch. Sunlight on snow or whitewash looks yellowish; and the shadows bluish. But the light from the lit and shaded areas differs only in intensity; its composition is the same. Even more impressive coloured shadows can be obtained in theatre lighting (see page 141). In the same way, two patches of, say, blue and yellow, often look brighter, and more contrasting, when adjacent than when apart.

We know from studies of the electrical activity of individual brain cells that a cell which is receiving no stimulus is dormant; but it is not totally inactive. In the absence of any message from the optic nerve, it discharges steadily: when the retina is stimulated by a flash of light of a particular colour, the electrical activity of the cell may be greatly increased, or reduced to almost nothing, or totally unaffected (see Figure 55). Different cells respond differently to different wavelengths.

Some brain cells appear to respond to colour only if it occurs at the edge of an area; such cells must play a crucial part in our perception of contrast, and hence of shape.

The boundary between two colours produces interesting effects. A patch of bright colour on a dark background appears, by contrast, brighter at the edges; a band of orange on deep green or brown appears almost to glow at the edges, like a strip-light, in much the same way as a series of wide stripes of increasingly paler grey seem, by contrast, to darken each of the boundaries. But, at the precise junction between two colours of similar

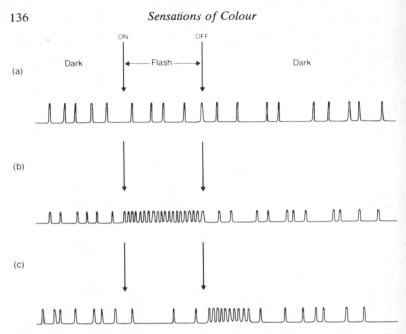

Figure 55. Brain waves *Some types of electrical activity of individual brain cells subjected to a bright flash of light.*
(a) No response: the cell continues to 'tick over', as in the dark.
(b) Activity increases during flash.
(c) Activity decreases during flash, but increases above normal level immediately afterwards.
Each cell may react differently to light of different wavelengths, and differently from its neighbours for light of the same wavelength.

brightness, the colours may become slightly paler, and so the contrast decreases. The effect is most marked when the two colours are complementary. For example, when yellow and blue stimulate adjacent cones, the response is much the same as if both cones had been triggered off by white light. The yellow and blue mix, not outside the eye, but on the retina, so each of the patches looks paler at the junction between them than it does further into the main area of colour. This effect obviously becomes more important the smaller the patches of colour. As early as 1824, Chevreul warned tapestry makers to avoid placing complementary colours next to each other if they wanted to produce brightly coloured pictures. For the same reason, painters and stained-glass artists often outline areas of colour in black, or separate them by a white line. But optical mixing need not yield only white; a mosaic of red and green gives a vibrant yellow. Such effects have been much exploited in textile design (Chapter 17) and painting (Chapter 19) and are the basis of colour television (Chapter 16).

Colours also become mixed on the retina if they follow each other more rapidly than about fifty stimuli per second. If the segments of a top are painted, the colours fuse when the top is spun (see Figure 56). Alternating slices of red and green become yellow, while blue and yellow become whitish, just as if lights were superimposed on a screen. A quite different effect, probably caused by different rates of recovery from fatigue to different wavelengths, is the stroboscopic colour we see when we spin certain patterned black and white discs (see Figure 57). The actual colours experienced depend on the observer, and vary with the speed and the direction of rotation, the particular pattern and the light source. The top shown here, designed by Sydney Harry, is at its most impressive when viewed by the light of a colour television set. Under fluorescent lighting it shows weaker colours, and under tungsten, or in daylight, none at all.

Colour sensations may change with time if one or more sets of cone cell become tired, as when an observer is shown a series of flashes of very intense red light (620 nm). The colour first seems red, then passes through orange and yellow to a green sensation, which holds for about thirty seconds before reverting gradually through yellow to orangeish-yellow, where it then stays. It seems that the barrage of high-wavelength photons is too great to be monitored for long by cones containing pigment A. Once the response is triggered, the recovery time is too slow to maintain the appropriate response ratio A:B:C for 620 nm. So a higher proportion of photons is absorbed by pigment B and the colour turns to green. Then the B cones also become fatigued, but, since they absorb a lower fraction of the photons than do the A cones, they become slightly less tired. So, after about two and a half minutes, the response settles down to give an orange-yellow sensation derived from a lower ratio of A to B responses than would be expected for 620 nm.

Fatigue also seems to account for the engaging apparitions known as 'negative after images'. If we stare hard at a coloured pencil held against a white background for one or two minutes, and then remove the pencil, we can see a pencil-shaped patch of complementary colour which moves across the background as we move our eyes; a red pencil gives a turquoise patch and a green pencil a magenta patch. It seems that, when we look at a red pencil, the cones respond, and become temporarily out of action, in the A:B:C ratio appropriate to the wavelength. When white light falls on the retina, these fatigued cones do not respond, and the light from the background produces a 'white' response in the unfatigued cones, but a 'white-minus-red' (i.e. turquoise) response in the pencil-shaped region of the fatigued cones. After a minute or so, the cones become reactivated and the

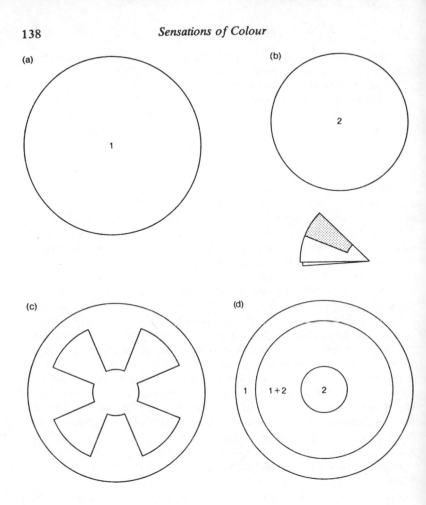

Figure 56. Coloured tops *Do-it-yourself colour mixing.*
(a) Circle of coloured paper (1) mounted on circle of stout cardboard.
(b) Smaller circle of paper of different colour (2), folded as shown, with shaded region cut away.
(c) Cut-out glued concentrically on coloured circle.
(d) When the cardboard circle is mounted (securely) on a child's top, or on the circular sanding base of an electric drill, colours 1 and 2 are mixed in equal proportions when the disc spins.
Best results are obtained with the brightest colours: a combination of fluorescent red and fluorescent green is particularly impressive.

Figure 57. Sydney Harry's top. *The original is 14.3 centimetres in diameter, and mounted on heavy cardboard with a sharpened matchstick as axle. It is spun by hand.*

after-image fades. Doctors and nurses wear bluish-green clothes when they do surgical operations so that they are not distracted by the bluish-green after-images of blood.

But not all negative after-images are as simple as these, or as Goethe's 'well-favoured' tavern wench (page 106). In 1965, it was found that if the eyes are 'primed' by looking alternately at green and black stripes in one direction (say vertical) and red and black stripes in another (say horizontal), then the negative after-images which are obtained when we look at black stripes on a white background are linked to the direction of the stripes. If we look at vertical stripes, we now see magenta; and if we turn the pattern through a right angle, so that the bars become horizontal, the colour changes to green. The connection between direction and colour can persist for hours, or even days, without impairing every-day vision in any way.

The exact explanation of this effect has not yet been fully worked out: but it is believed to hold the key to a greatly increased understanding of our perception of colour, and of the nature of simple learning and memory. One thing is certain. Much more is involved than the mere time which it takes a cone to recover from absorbing a photon.

Negative after-images, of colour complementary to the object, are normally obtained when the illumination is fairly even. Incandescent sources may produce a positive after-image, a fleeting copy of the object. In a darkened room, we may continue to see a candle flame for a moment or two after it has been blown out. And, if you have a short enough name, you may write your whole signature across the night sky with a lighted sparkler, for the colour lingers for an instant after the sparkler has moved on.

Perhaps the most impressive of all after-images is the 'flight of colours' which we see after a very bright light has been shone into the eye, as when we have been examined by an oculist. Against a white ceiling floats a brilliant yellow patch with a magenta border, edged with turquoise around its changing amoeboid shape. The colours gradually change too, so that it becomes a negative after-image: blue, edged with yellow and red. The effect lasts for several minutes, changing shape and colour like a spreading oil patch.

Visual information can also be stored for much longer periods. We can remember some colours for years, and our memory can modify the colours we now see. If two shapes, representing a leaf and a donkey, are cut from the same piece of grey paper and placed on the same background, most observers claim that the leaf is a greener grey than the donkey.

A similar expectation may be one reason why we say that a donkey, lit by the setting sun, is brownish grey, although the light which reaches us from it is akin to that scattered by rust or copper in daylight. But the state of adaptation of the eye is also important. It is thought that the retina can adapt to wavelength composition of the illumination, if this differs appreciably from white light, in much the same way as it adapts to changes in intensity. Those who use cars with slightly tinted green glass will know that, seen through the windscreen or a closed window, the surroundings look 'normal' and the glass untinted. The eye is adapted to a faintly green background. But if we then partly wind down a window, that part of the scene we see through the open window now looks distinctly pink, while the adjacent part, viewed through the glass, is tinged with green. After a minute or so, however, the contrast fades, and both parts of the scene look untinted.

More dramatic illustrations of the interplay of adaptation and contrast can often be seen in theatre lighting. The stage is lit from two widely separated spotlights, one white and one red, so that their spots exactly coincide on the white background, which our eyes, once they have adapted to the pink light, see as white. A dancer appears, and blocks the beam from the white spot; and where her shadow falls on the background, the only light which reaches us is from the red spotlight. So we see a red shadow on a white background. Her partner then emerges from the opposite wings, and blocks the beam from the red light. So where he casts a shadow, only white light reaches the background. But we do not see his shadow as white on a pink background, because we already see the background as white. So, instead of perceiving it as pink-minus-red, we see it as white-minus-red: as turquoise. The shadows of the *pas de deux* appear, not as red and white on pink, but, more impressively, as red and turquoise on white.

A similar effect has been exploited by Land, who showed that red and white light could be mixed in such a way as to give sensations over a wide range of colours. He made black and white positive transparencies, photographing each scene twice, once through a red filter, and once through a green one. He projected them both on to a screen, the first with a red light, and the second with a white one, and when the images were superimposed, obtained, in addition to red and black, a range of colours from turquoise, through green and yellow, to orange.

We do not yet know if the mechanism by which we adapt to the colour of the illumination takes place in the eye or in the brain. But there seems to be a series of brain cells, each of which picks up processed signals from a very narrow band of wavelengths. Between them, they span the visible spectrum; and it seems that each discernible colour (including the non-spectral purple) is associated with its own set of brain cells. Of these cells, a number will respond only in the presence of the background illumination. Thus some of the cells which respond to changes in the intensity of red light (but are insensitive to changes in the intensity of other colours) will respond to red *only* if it is mixed with light of other colours; pure red light produces no activity in such cells. It may well be that observations of this type will eventually help us to understand why the perceived colours of objects depend so little on the light which falls on them.

But the extent to which the colour of an object seems to remain constant also depends on who is looking at what. The colour of a real donkey seems to change less with the illumination than does the colour of a donkey in a picture. The colours of a naturalistic painting vary less than do those of a bold pattern, which are themselves more constant than an abstract picture with ill-defined regions of colour. To a large extent, what we see depends on what we expect to see. Even to an artist, a donkey may, at a brief, casual glance, look donkey-coloured; but if the artist is thinking about incorporating the donkey into a painting, he will be more sensitive to the light reaching him from the donkey at that particular moment than to the average colour which the donkey would have if viewed by a layman in diffuse light.

It is not surprising that the ability to look analytically can be cultivated by willpower and training; nor that those without training often prefer a naturalistic picture which represents colours as they remember them to those in which the artist tries to reproduce the actual light coming from the object. In much the same way, an observer will often select, as the most 'natural' colour photograph, one in which the colours are actually brighter than those in the original scene. Colours selected to match memories from dreams, however, are often paler than those of similar objects we observe when awake.

Colour sensations may often be even more dramatically altered by, or even produced by, factors other than light. Some forms of hysteria and hallucinogenic drugs enhance appreciation of colour: hallucinogens often provide experience of brightly coloured patterns which bear little relation to the real world, and those about to suffer attacks of epilepsy or migraine may see coloured rays and geometrical shapes before an attack. The blind, particularly if they are old, may have similar sensations, akin to seeing 'golden rain' and coloured patterns. In childhood, a pressure on a closed eye was enough to produce patterns as exotic as Catherine wheels or peacock tails. In adult life, a firmer touch, preferably on the upper part of the eye, is needed to produce an inferior but none the less impressive result. Electrical and mechanical stimulation of the optic nerve and the visual areas of the brain can make us see colours; so can some acute illnesses, and even a strong magnetic field. And so, of course, can memory and dreams. But although we can experience an immense variety of colour sensations produced in these different, non-visual, ways, we have as yet only a negligible understanding of any of the mechanisms involved.

15

SORTING AND RECORDING

To what extent can we record a colour? Can we impose any order on our rich variety of colour sensations? If we can, would our scheme be entirely personal, or could we use it to communicate information about a colour? How can we best tell someone the exact colour we should like the new door to be?

As we shall see in Chapter 21, there are many difficulties in trying to describe colours with words. A request to paint the door turquoise would be likely to produce a fairly bright door in the greenish blue (or might it be bluish-green?) range, neither very pastel nor very murky. Perhaps we could get nearer to the colour we have in mind by a request to paint the wall the same colour as the curtains. But, though the two may seem a good match in one light, they may clash horribly in another (see page 152). And as they will have different textures, they will have different highlights; so although the colours of the two may 'go' very satisfactorily, they will certainly not match all over, even if the light falling on them happens to be identical. It is safer to choose from samples of the actual paint, on the manufacturer's own colour card. But maybe the exact colour required is not available. 'Please mix me...' How does one continue? 'Something between these two'; 'Something like this, only lighter'; 'A more subtle shade of that'; 'A bluer version of this one.' It is difficult to know how to specify, or how to produce, the colour required. And it may even be difficult to envisage precisely any colours which are not on the colour card.

Perhaps it would help if we could arrange colours in some rational order and label them appropriately. We could then refer to them in much the same way as we can pinpoint a place by a map reference. Many child-hours must be passed in just this way, rearranging crayons, pastels or embroidery threads. It is easy enough to make a line of the rainbow colours, and join the ends, through purple, to give a circle. But problems soon arise. Do black, grey and white count as colours? If they are to be included, where should they go? What should be done with pale colours: primrose, duck-egg blue, salmon? And what about the browns? One can imagine a cross-roads at yellow, with primrose leading to white on one side, and ochre leading to khaki, brown and black on the other. Perhaps, however, black and white should be joined through the greys, to give another circle, perpendicular to the first. It seems we cannot arrange our coloured objects on a flat surface, but need some three-dimensional scheme.

Maybe it would be better to seek a more 'scientific' classification than any subjective arrangement of coloured materials? We might try to specify the colour of a sample by irradiating it with the light of a large number of very narrow bands of wavelength and measuring the percentage of each which the sample reflects. These measurements can be made extremely easily, given the appropriate equipment. But the light which enters the eye depends on the lighting as well as on the sample, so we would also need to know the composition of the illumination. Even this does not tell us the colour of the sample unless we know how the eye reacts to light of different wavelengths. Two materials may match exactly under one type of illumination even if they send light of totally different composition to the eye; we know that many yellows can be matched by mixtures of red and green light. We might, however, combine, for each narrow band of wavelength, measurements of the reflecting powers of the sample, and the composition of the light source with our knowledge of the response of the retina. Although this procedure still needs laboratory equipment, it relates the scientific measurements of the light reaching us to our perceptions of colour, assuming that the observer has normal colour vision, adapted for daylight viewing and uninfluenced by the effects (such as after-images, contrast of near-by colours, memory, expectation) which we discussed in the previous chapter. So we are attempting to chart, not just the stimulus of the light entering the eye, but a normal observer's response to it. The idea of attempting to *measure* colour in this way sounds attractive, if somewhat complex. But do its advantages always outweigh those of map references with an ordered arrangement, albeit subjectively chosen and represented in three dimensions? Since both systems are used in practice, we shall look at each in more detail.

If we are to arrange a number of colours in any systematic order, we must decide what qualities we shall use to sort them. Let us first recall the ways in which light can vary. The sensation of colour depends primarily on the composition of the light, and partly on the intensity; and the composition may be usually described as a mixture, in a certain proportion, of white light with a 'coloured' light of a particular dominant wavelength within the visible spectrum. (For purple light, the 'coloured' component is itself a mixture.) We can describe the primary sensation of colour in terms of *hue*, which refers to the greenness, blueness and so forth, and varies with any change in the dominant wavelength. The extent to which this wavelength in fact dominates the light is known as *saturation* (or chroma). As the dominant wavelength is diluted with white light, the saturation decreases. An increase in intensity in light of a particular composition increases the *brightness* (or value). The so-called 'natural' or 'achromatic' colours, black, grey and white, are of zero saturation, and differ from each other only in brightness. A series obtained by adding one hue, say blue, to white differs only in saturation, as does a series obtained by adding a blue pigment to a grey one. If the same pigment were added to a white one in the same proportions, the dusky, pale blue would differ from the clear pale blue only in brightness.

There are many three-dimensional arrangements of colours, the best known being those devised by Munsell and by Ostwald. Both are based on the colour circle formed by joining the two ends of the spectrum through purple (see Figure 58). So the hue changes around the circumference of the circle, much as the hours progress around the face of a clock. Through the centre of the clock face, and perpendicular to it, like the axle of a wheel, runs the line representing the neutral colours, usually with white at the top, changing, through deepening greys, to black at the bottom. Radially, like spokes on a wheel, the saturation increases towards the rim, to give a space which can be filled in by different colours, according to which system is being used.

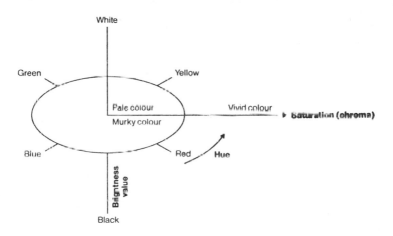

Figure 58. Colours in space *The skeleton of a colour solid. The 'achromatic' colours form the vertical backbone from which the different hues radiate: red in one direction, green in the opposite one, and the others in between. For any one hue, the colour becomes more vivid the farther it is from the centre.*

Ostwald arranged his colours in a double cone, based on twenty-four different hues, arranged around the circumference (see Figure 59). Each hue is combined, in a number of fixed proportions, with each of eight

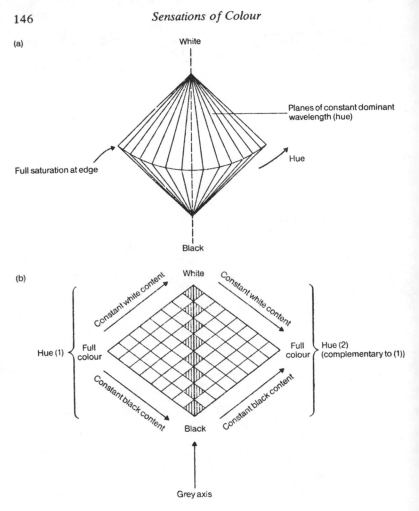

Figure 59. Ostwald's colour solid *(a) Exterior view, (b) Vertical section.*
(Adapted, with permission, from G. J. Chamberlin and D. G. Chamberlin, Colour: Its
Measurement, Computation and Application, *Heyden, London, 1980.)*

equally spaced neutral colours from white to black. The resulting colours
are arranged so that brightness decreases vertically towards the bottom of
the diagram, while saturation decreases towards the centre. Thus Ostwald's
colour solid consists of twenty-four triangles (one for each hue), arranged
radially so that a vertical section through it gives two such triangles, for
complementary hues, fused at the centre. Each position within the solid is

numbered on a grid system, so that any colour contained by the solid can be specified by a map-reference.

In Munsell's arrangement, saturation is increased by a series of visually equal steps rather than by adding a fixed proportion of pigment; and there are nine neutral colours, rather than eight. As the number of equal steps of saturation at a particular hue and brightness depends on the hue, the arms are of different length in different parts of the solid. Munsell's solid is therefore much less regular than Ostwald's, and on account of its untidy appearance is known as a colour 'tree'. A typical vertical section through it is shown in Figure 60. Each position in the tree, as in Ostwald's cone, is encoded, thereby allowing colours to be specified. And the tree has one great advantage over Ostwald's solid: whenever some new, dazzling pigment is made, it may be incorporated by extending an existing branch.

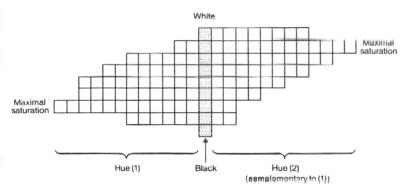

Figure 60. Munsell's tree *Vertical section (cf. Figures 59(b), page 146, and 83, page 226).*

The two systems resemble each other in that the circumference is divided arbitrarily into hues, and the vertical axis is graduated into visually equal steps, which are obtained by asking large numbers of observers to estimate equal differences in brightness. But the two solids are based on different ways of varying saturation. While Ostwald used the ratio of neutral pigment to saturated pigment, Munsell again invoked visual assessment by the average observer.

When Ostwald devised his colour solid, he was an old man whose sensitivity to blue was doubtless declining, which accounts for the slight compression in the blue region of the circumference. Swedish workers have attempted to correct this defect in Ostwald's system by placing each of the four psychological primaries at right angles to one another on the

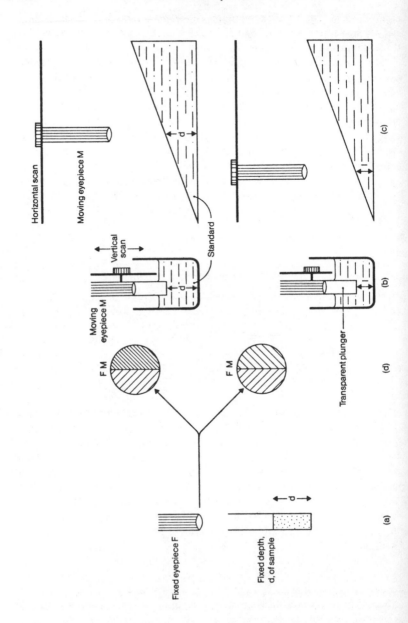

circumference and dividing the segments between them into visually equal steps, constituting a more 'natural' colour circle for use as the basis of a three-dimensional colour solid.

Although a colour solid is a useful concept, and may even be constructed as a display object of great visual and intellectual appeal, either swatches or books are more convenient for everyday use. Colour atlases, such as the *Munsell Book of Colour*, often represent vertical sections of a colour solid, cut through each hue represented. For more specialist use, a restricted range of colours, varying by smaller gradations, may be reproduced as in collections for those who wish to specify the precise colour of a rose petal or a sample of human skin or tooth.

Such visual matching of colour is a stepwise process, a placing of the sample of unknown specifications between two standard colours whose specifications are known. But how can we try to measure the colour specifications of a material if we have no standard colour which matches it? What can we do to try to measure 'colour', to produce specifications of hue, saturation and brightness?

Since colour is a sensation, there is a lot to be said for the measurements being made by the eye. The human eye is, in fact, an excellent detector of differences of hue, and many people can assess the percentage of red, blue and yellow in a pigment with surprising precision. But human estimates of saturation, and of brightness, are much less reliable.

The most precise visual methods of attempting to specify colours, like the use of a colour atlas, involve matching. The simplest are those devised merely to measure saturation, as in the determination of the concentration of a single coloured component in a liquid (see Figure 61).

If the two solutions appear to be the same colour, the ratio of their concentrations is simply related to the ratio of the length of the two solutions through which the light has passed; and this may easily be found using a simple comparator involving either a plunger or a wedge.

Figure 61. Comparison of saturation *A fixed depth of the sample is viewed with one eye and a variable length of standard with the other. The geometry of the instrument is adjusted until the same depth of colour is observed by each eye.*
(a) Sample.
(b) Variation of depth of standard by means of transparent plunger.
(c) Variation of depth of standard by use of wedge.
(d) Split field, one half from each eyepiece.
In the top two diagrams, the same depth used for sample and standard gives a darker field on the right. In the lower diagram, the length (l) of standard has been adjusted so that the two fields are indistinguishable. In this example, d = 2l, indicating that the saturation of the sample is half that of the standard.

For many purposes, however, we need to know the hue, as well as the saturation. We again compare our sample with a colour which we can specify. But how can we vary this colour whilst still being able to specify it? One way is to mix coloured lights of known wavelength in known proportions. A simple arrangement is shown in Figure 62. The amount of red, blue and green light can be varied by horizontal and vertical movement of the filter assembly over the source of light, and the three lights are then mixed by diffusion and multiple reflections. More sophisticated devices, used mainly for research on colour, involve six lights of varying wavelengths. In each case the mixture of coloured lights, shone on to a white background, is matched with the unknown sample, illuminated from a standard source of white light.

Alternatively, the sample may be matched by a patch of coloured light which has been obtained by passing white light through three filters, one

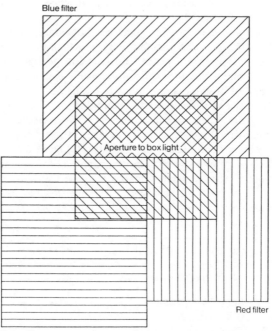

Figure 62. Mixing lights *The required mixture of red, blue and green light may be obtained by adjusting that area of each filter which lies over the aperture to the mixing box.*

magenta, one yellow and one blue-green (see Figure 63). Sets of such filters are available commercially for use in an instrument equipped with a standard light source and known as the Lovibond Tintometer. The full range of 250 filters of different depth for each of the three hues allows nearly nine million different colours to be obtained, including the full range of achromatic colours from white to black. The colour of the sample is readily specified in terms of the three filters used to match it.

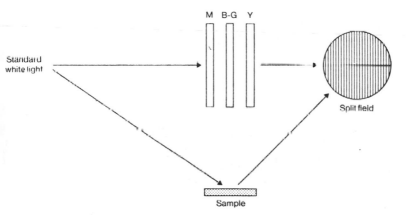

Figure 63. Labelling with filters *The colour of the sample can be matched with that of light which has passed through three Lovibond filters, magenta (M), blue-green (BG) and yellow (Y) of specified strength. A purplish blue sample, for example, would need a deep magenta (to absorb most of the green), a medium blue-green (to absorb some, but not all, of the red), and yellow of appropriate depth to reduce the intensity of the colour to that of the sample.*

Instead of matching the light which reaches us from a sample with that from an unknown, we can exploit the phenomenon of persistence of vision and match only the sensations. Split discs, coloured in saturated blue, green and red, are placed on a revolving platform in such a way that the proportions of the three colours can be varied (see Figure 64). When the platform is spun at high speed, the sensations merge, just as if three coloured lights were superimposed. The areas of the three colours are adjusted until the colour of the spinning disc exactly matches that of the sample, which can then be expressed in terms of the proportions of primary colours used.

Nowadays, there is an increasing tendency to use photoelectric instruments for colour matching, instead of the eye of one or two individual observers. Such instruments monitor narrow bands over the whole visible spectrum, first detecting how much light in each band reaches them from the sample, and then converting this information into the size of the stimulus which bombards each of the three cone systems in a normal human eye. Finally, the responses to these three stimuli are combined to give the colour experienced by the 'standard observer'. Inescapably, these instruments give results based on sensations experienced as a consequence of human vision. But the vision is that, not of a few individuals, but of the 'standard observer' built up from observations made by a large number of individuals selected for their normality of vision. But even when the matching is done, rather approximately, by eye, it is seldom left for a single individual: more often two, or even three, observers are used.

(a)   (b)

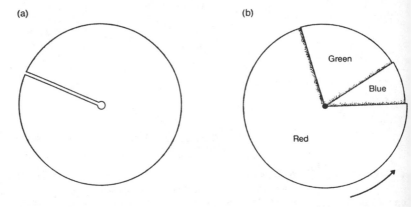

Figure 64. Tops for colour measurement *(a) Split circular disc of paper of standard colour. (b) Standard green, blue and red interlocked, exposing known areas of each for colour mixing when the disc is rotated.*

Matching, whether by eye or machine, often gives different results with different sources of light; and as matching implies identity only of *response*, this is no surprise. Two extreme examples of identical colours produced by light of very different composition were given in Table 3 (page 119). Imagine a pair of yellow pigments, one reflecting light of only 580 nm and one reflecting only light of 540 nm and 630 nm, each at half the intensity of the first pigment. If both were illuminated with light containing equal intensities of the three wavelengths, the two pigments would match exactly. But if they were illuminated with light containing a slightly higher proportion of longer wavelengths, the second pigment would look redder than the first. Such 'metameric' pigments, which match only under one light source, are the bane of those who try to match clothes and accessories, or paint and fabrics (see Figure 65). Metamerism accounts for much of the popularity of 'colour coordinated' goods sold by the same manufacturer who uses dyes such that any change in illumination causes almost the same change in colour for the different materials.

Colour solids and atlases can give us no information about the composition of the light which causes a particular colour. For charts relating colour

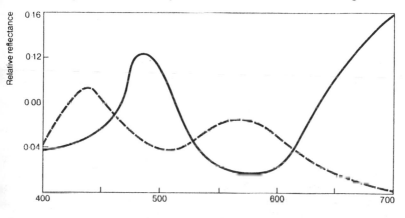

Wavelength (nm)

Figure 65. A good match? *The reflectance spectra of two fabrics which match perfectly in daylight. (Reproduced, with permission, from W. D. Wright,* The Measurement of Colour, *4th edition, Adam Hilger, London, 1969.)*

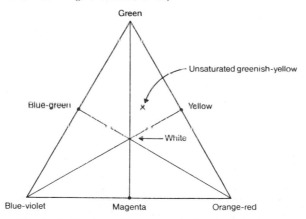

Figure 66. Maxwell's triangle *Shows how many, but not all, colours can be represented as a mixture of three primary coloured lights. The nearer a point is to an apex of the triangle the higher is the proportion of light of the colour represented by that apex. The point X (50 per cent green, 35 per cent orange-red and 15 per cent blue-violet) represents an unsaturated greenish yellow.*

to composition, we turn to the second, more 'scientific' approach. Just as many colour solids are based on a ring of spectral colours, joined through purple, a triangle usually forms the basis of attempts to chart the colours produced by the mixing of lights. As early as 1855, Maxwell found that a great number of colours could be produced by mixing lights of only the three 'primary' colours: orange-red, green and blue-violet. The colour resulting from a particular mixture can be represented by a point on a triangular grid (see Figure 66). Many colours can be specified in this way: but not all. Whichever three primary sources we choose, there are always some colours (including many pure spectral ones) which cannot be represented by a point in, or on, the triangle; which confirms that we cannot always match one colour by a mixture of three others unless we allow ourselves the option of mixing one of the primary colours with the sample and then matching the result with a mixture of the two other lights. Figure 67 gives the recipe for obtaining a match for every visible wavelength with three primaries. Thus vivid yellow (570 nm) (cf. page 119) can never be exactly matched by red (700 nm) and green (546 nm); but if a little blue (436 nm) is added to the yellow, a perfect match can be made. We can express this algebraically by stating that vivid yellow can be matched by red, green and a small negative amount of blue. But since there is no scope for plotting negative contributions on a Maxwell triangle, colours such as vivid yellow cannot be represented on it.

It is too bad that we cannot choose any three wavelengths which, when themselves mixed together, will produce *all* visible colours. But there is nothing to stop us imagining that such ideal primary colours might exist; and if they did, they could be mixed in such a way as to produce three convenient real primaries, such as Maxwell used. So we could draw a mathematical modification of Maxwell's triangle, with our three imaginary primaries at the corners. Points for the real ones, and for all other colours, would then be within it. Three such imaginary primaries have indeed been devised, such that any real colours can be represented as a mixture of the appropriate amounts of the three of them, and plotted as an idealized version of the Maxwell triangle. But not everyone is used to triangular graphs and the ruled paper may not be easy to obtain. Could we not use squared paper instead? How many variables do we need? Since we have decreed that a real colour R can be matched by a mixture of our three Imaginary Primary Lights – say, I units of one, P units of the second and L of the third – it might look as though we need the three variables, I, P, and L, to specify R. But we could also say that the brightness of the light is the sum S of the three primaries (so $S = I + P + L$), mixed in the ratio of $I/S : P/S : L/S$. And if we know I/S and P/S we already know L/S, because $I/S + P/S + L/S = 1$. Since we are often more interested in colour than in

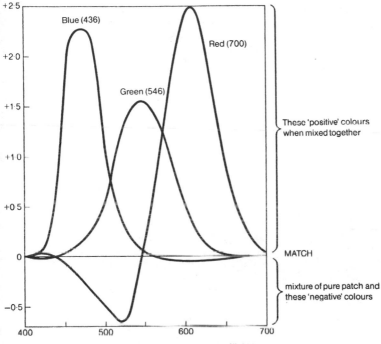

Figure 67, 'Negative' colours *Many pure spectral colours can be exactly matched with three primaries only by use of 'negative' colours. When one primary is mixed with the 'pure' patch of light, this mixture can be matched by some mixture of the other two. For the primary lights used here, the only colour which can be matched directly by the three primaries is a greenish yellow in the region of 550 nm. (Reproduced, with permission, from F. W. Billmeyer and M. Saltzman,* Principles of Color Technology, *Interscience, New York, 1966, p. 33.)*

brightness, we could concentrate on the two quantities I/S and P/S which specify the colour. We could then use ordinary squared graph paper. If we ever needed to know L/S, it would be easy enough to calculate it. And if we decided that, after all, we wanted to specify the brightness, we could represent it on an axis rising vertically, out of the paper. The final diagram would be much like a plan, with the two specifications of colour running north-south and east-west; or, if brightness is added, like a map which also shows heights, or isobars (or any third variable) superimposed on the plan as a series of contours.

So all that we need in order to specify the colour of an object is a knowledge of the composition of the light falling on it, of the modification of the light by it, of the response of the normal human eye, and of the quantities of three imaginary primary lights which would produce the same response; and a piece of ordinary graph paper. The Commission Internationale de l'Eclairage (CIE), in 1931, defined the standard observer and three possible standard sources; and they produced tables showing the relationship between the observers' response and the quantities of the imaginary primary lights which would in theory be needed to produce them. The tongue-shaped curve (see Figure 68) in the graph shows the specifica-

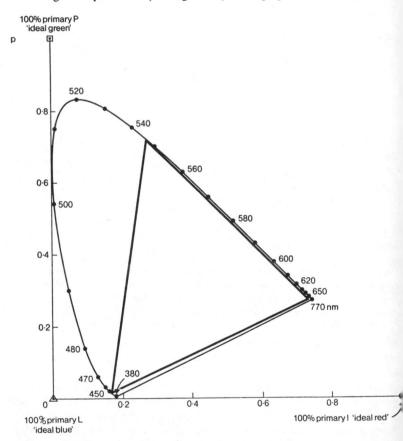

Figure 68. CIE tongue diagram *The tongue encloses all visible colours, with the pure spectral ones lying along its curved edge. The inner triangle encloses those colours obtainable by mixing real primaries 436 nm, 546 nm, 700 nm (i.e. those enclosed in Maxwell's triangle of Figure 66, page 153).*

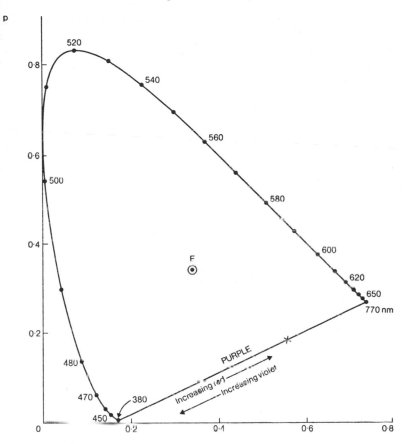

Figure 69. Purple and white *The point E for p = 0·33, i = 0·33 (and so l =$\frac{L}{S}$= 0·33) is 'equal energy' white. Purples lie along the base line (see text).*

tions of the pure spectral colours as their values of i=I/S and p=P/S. (Modified definitions of the standard observer and the standard sources, introduced by the CIE in 1967, change only details on the graph.) The diagram, known as a CIE chromaticity curve, has been used to depict colours, and the relationship between them, in a wide variety of situations.

The enormous usefulness of the CIE diagrams arises from the fact that we can represent a mixture of two lights as a point on the line joining the points which specify them. So purples, formed by mixing red and blue, lie on the base-line of the 'tongue' (see Figure 69). A mixture of two parts red

(770 nm) and one part violet (380 nm) would lie at point X, twice as *near* to the red point as to the violet point. Since all the colours formed by mixing real lights lie inside the area enclosed by the tongue, it is only those colours which are represented by points inside the curve which are visible. As the area outside the curve represents only imaginary stimuli, we need not consider it further.

We can also use CIE diagrams to specify a colour in terms of its dominant wavelength and its saturation, as well as in terms of the contributions of responses to imaginary primaries. The point E on Figure 69 represents an equal mixture of the three primaries and hence its position is $p=0\cdot\dot{3}$, $i=0\cdot\dot{3}$ (and so $l=0\cdot\dot{3}$). It is known as equal energy white. Suppose now that we want to find the dominant wavelength of the colour represented by point X in Figure 70a. We can think of this colour as being some mixture of a

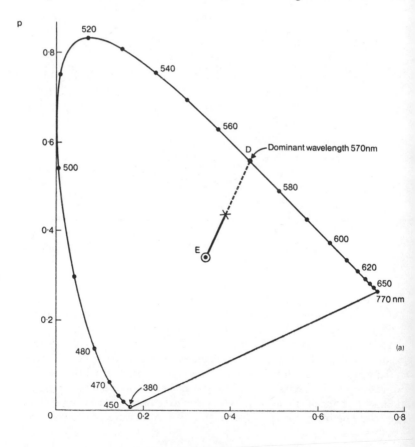

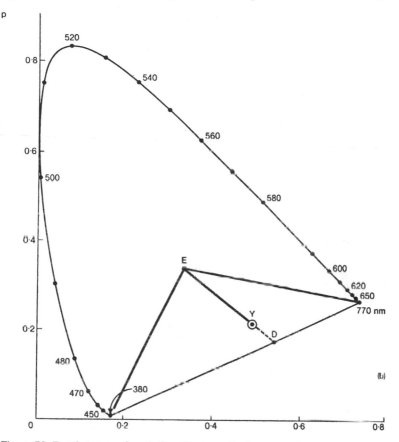

Figure 70. Dominant wavelength *Specification of colour as a mixture of white light with;*

(a) *One spectral wavelength.*
(b) *A non-spectral mixture.*

(See text.)

dominant wavelength D with white light of composition represented by E. To find which is the dominant wavelength, we remember that X must lie on a line joining E and D. So we draw a line from E to X and continue it until it meets the curve, at the dominant wavelength D. So X is an unsaturated version of the pure spectral colour D. How unsaturated? We know this from the distance between the mixture X and the dominant wavelength D. In this example, X is exactly half-way between pure colour and pure white and so can be specified as a 50-50 mixture of light of 570 nm and white light.

What happens when the line joining E and the point for a colour does not meet the curve of pure spectral colours, but cuts the purple base line? For point Y in Figure 70b, the 'dominant colour' is point D, on the line joining the blue and red ends of the curve. Since D is half as far from the red end as from the blue end, it represents a purple mixture of two parts red to one part blue. The colour Y can be obtained by mixing this purple with white; and as the distance between white and Y is three times that between Y and the dominant colour, Y is a mixture of three parts of D with one part of white.

We can use CIE diagrams, not only to represent colours, but also both to

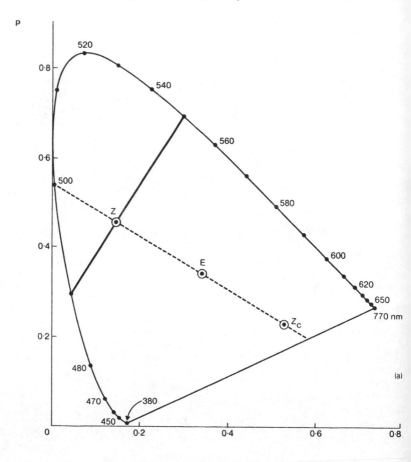

(a)

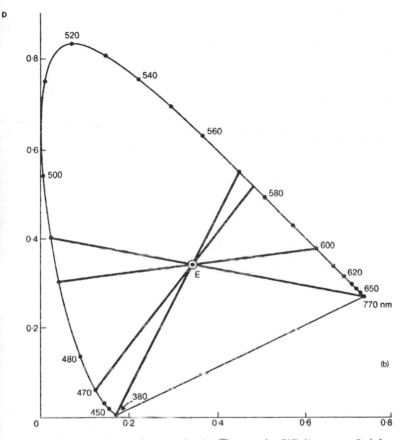

Figure 71. Colours and complementaries *(a) The use of a CIE diagram to find the colour of a mixture and its complement (b) Pairs of complementary pure spectral colours (see text).*

predict the result of mixing coloured lights and to locate complementary colours. Suppose we mix four parts of 490 nm with three parts of 550 nm; we obtain a colour represented by point Z in Figure 71a. But what colour is it? We find out, as before, by drawing a line from E, to Z, and beyond, till it cuts the curve, at 500 nm. So this is the dominant wavelength. The distances of Z from pure colour and pure white tell us that Z could be made from six parts of 500 nm and five of pure white. And the complementary colour? This is the colour Z_c, which, when mixed with Z, gives white. So Z, Z_c and E must lie on the same line; and the points Z and Z_c must be the same distance from E. We can therefore find Z_c by extending the line from Z to E by an equal distance beyond E. The colour complementary to point Z (unsaturated green) is unsaturated reddish purple.

The colour Y_c which is complementary to a pure (saturated) spectral colour Y is itself saturated and can easily be located by drawing a line from the point Y through E and beyond until it cuts the other side of the curve at the required point Y_c (see Figure 71b).

So far, we have ignored any change in brightness, even though we could add this information to the CIE diagram by using a vertical scale. We know (page 120) that the brightness of a light depends both on its intensity and on its wavelength, since the eye is more sensitive to some wavelengths than to others. So the brightness of an object depends both on the object itself and

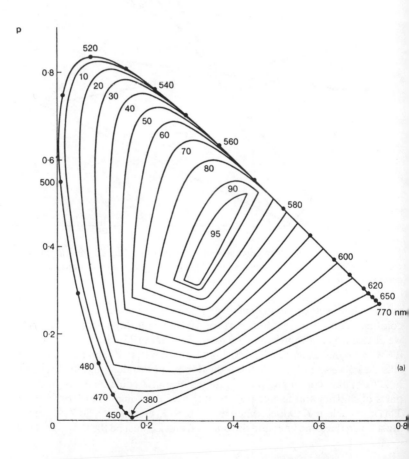

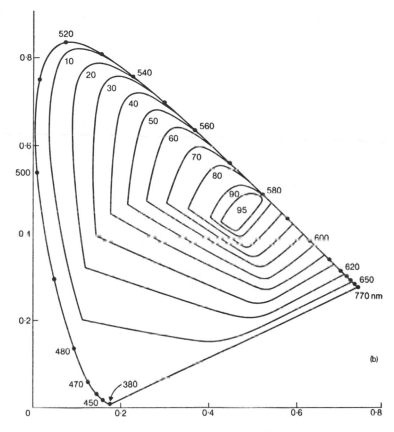

Figure 72. Brightness *Maximal brightness contours of colours in:*
(a) Daylight.
(b) Tungsten light. (Reproduced, with permission, from F. W. Billmeyer and M. Saltzman, Principles of Color Technology, *Interscience, New York, 1966.)*

the composition of the light falling on it. Strong colours (except for yellow) tend to be dark; and light colours tend to be unsaturated. It follows that when the brightness is high, there is rather little variation in colour. The limits within which colour can vary for a particular brightness lie within the contours shown in Figures 72a and b. As expected, very high brightnesses are restricted to the white and yellow regions of the diagrams. And, also as expected, the shapes of the contours are very sensitive to the composition of the illuminating light, maximum brightness being shifted from white in daylight to yellow in tungsten light, which contains a much lower proportion of blue.

Most CIE diagrams are, however, two-dimensional and ignore brightness. Colour is represented only by hue and saturation. Despite their esoteric basis and the conceptual unpalatability of imaginary lights, CIE diagrams are very widely used to give a compact and coherent display of some very varied information. A few examples are given in Figures 73 to 75. One disadvantage of ignoring brightness is that CIE diagrams, unlike colour solids, make no distinction between white and grey. All the achromatic colours, represented separately up the vertical axis of a colour

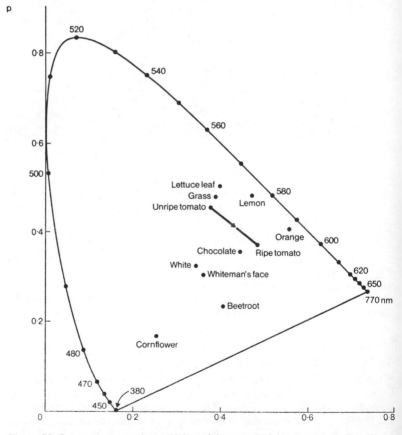

Figure 73. Some common colours *(Adapted from A. Padgham and J. E. Saunders, The Perception of Light and Colour, G. Bell, London, 1975.)*

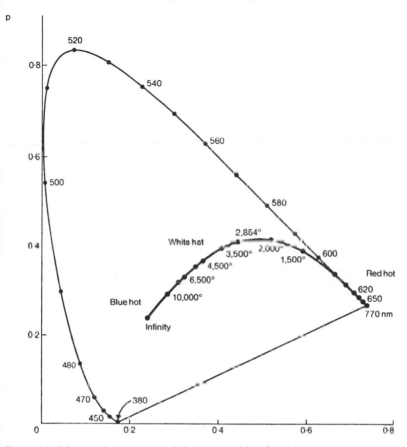

Figure 74. Colour and temperature *Colours emitted by a hot 'black' body (see page 56). (Adapted, with permission, from D. B. Judd and G. Wyszecki, Color in Business, Science and Industry, John Wiley, New York, 3rd edition, 1975.)*

solid, are plotted as a single point on the CIE tongue; since they are all composed of equal ratios of the three primary lights, they are all represented by the point E. And what of the browns, which feature in a sizeable slice of the colour solids, but seem to have disappeared from the CIE diagrams? These are now merely low-intensity yellows, flanked by low-intensity greens and oranges. On the CIE diagram, a mid-brown occupies the same point as a mid-yellow.

In this section, we have discussed some of the processes which contribute

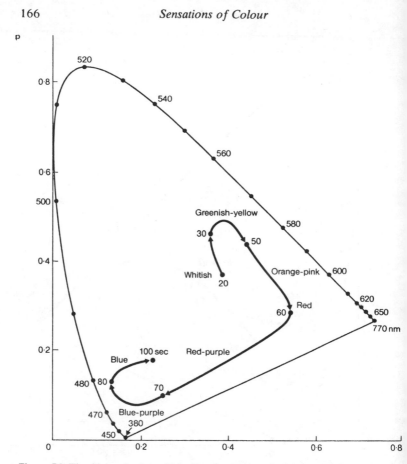

Figure 75. The flight of colours *How the after-image of a bright white light changes with time (see page 140). (Adapted from A. Padgham and J. E. Saunders,* The Perception of Light and Colour, *G. Bell, London, 1975.)*

to our sensations of colour, and some of the devices we use to try to record these sensations. The rest of the book deals with the way in which man controls these sensations: with the various contexts in which man makes use of colour.

PART FIVE:
TECHNOLOGY

That ... is the most praiseworthy which is most like the thing represented.

– LEONARDO DA VINCI

Colours fixed in glass by fusion, and by nature in gems, defy all time and re-action.

*

It is curious ... to take a glance at the works containing directions on the art of dyeing. As the Catholic, on entering his temple, sprinkles himself with holy water, and after bending the knee, proceeds perhaps to converse with his friends on his affairs, without any especial devotion; so all the treatises on dyeing begin with a respectful allusion to the accredited theory, without afterwards exhibiting a single trace of any principle deduced from this theory, or showing that it has thrown light on any part of the art, or that it offers any useful hints in furtherance of practical methods.

– GOETHE, *Theory of Colours*

Bright yellow was dyed by means of 'flowers of St John' (Anthemis tinctoria), for which the yarn is boiled with alum in a copper kettle, although not with too much alum, then it is dried; after that, the 'flowers of St John' are boiled for a long time in water and finally the mordanted yarn is added to the decoction ... *Greenish yellow* is dyed with birch leaves boiled in water after mordanting with alum. *Green* is dyed thus: first the yarn is dyed yellow with Saw-wort (Serratula tinctoria) and then dyed blue with *Indigo*, since yellow and blue together make green. *Sea-green* was dyed with diluted acetic acid and salt in an untinned copper kettle, that was left till it became verdigrised. Then the yarn was put into it and stirred often to make it evenly dyed, but it was not boiled.

– CARL LINNAEUS, *Öland and Gotland Journey* (1741)

It is also to be remarked that ladies, in wearing positive colours, are in danger of making a complexion which may not be very bright still less so, and thus to preserve a due balance with such brilliant accompaniments, they are induced to heighten their complexions artificially.

*

I remember a Hessian officer, returned from America, who had painted his face with the positive colours, in the manner of the Indians; a kind of completeness or due balance was thus produced, the effect of which was not disagreeable.

– GOETHE, *Theory of Colours*

16

COLOUR REPRODUCTION

Imagine an occasion of historic importance, a state visit, maybe, a royal funeral or the signing of a treaty. Reporters of all types would be out in force to record the scene for their current public and for posterity. There would be colour cameras, for television, for videotape, for the cinema and for colour prints and slides. Some of the still photographs would be destined for reproduction as book illustrations, posters and cards. And, in all these techniques, the aim is usually to make the colours in the picture as 'true' as possible to those in the original scene.

A colour reproduction, whether anchored on paper or flashed on a screen, must try to produce the same sensation as that produced by light from the original scene. The basic problem is theoretically the same as that involved in matching one small, homogeneous patch of light with another (see page 150); in selecting two identical samples of paint; or in choosing a ribbon to match a dress.

A real scene is, of course, rather more complicated than two patches of coloured light. In fact, a whole eyeful of scene can be broken down into 100,000 separate patches. Many people are able to distinguish at least thirty different 'pure' colours represented by the curved edge of the CIE diagram (say, bands of width 10 nm within the range 400 nm to 700 nm). And we can, of course, distinguish between a very much larger number of the non-pure colours represented by small areas inside the CIE curve, and by the samples in the various colour atlases. Estimates of the number of colours which can be discerned by people with good colour vision vary from about half a million up to as much as ten million. In theory, a perfectly matched picture could be produced if the light reaching each of the 100,000 patches on the retina was first analysed for wavelength composition and intensity, and then reproduced. But this task would require such technical and economic resources as to be effectively impossible. So another line of attack must be used.

All colour reproduction is based on the attempt to reproduce, not light of a definite composition, but colour; that is, the sensation which the light produces. Thus, the light of a yellow street lamp could equally well be represented by light of 500 nm or by a mixture of two lights of wavelengths 540 nm and 630 nm (see page 118). Although the processes used in different types of colour reproduction vary considerably, they are all based on the same idea: the light from each point in the scene is monitored in some way which is related to the effect of the light on the retina, and the information obtained from this analysis is used to produce light which will have a similar retinal response. Despite the fact that none of the techniques of colour reproduction has yet achieved a perfect copy, good matches can be obtained by using the basic procedure described below.

First, each point in the scene is scanned by three light-sensitive devices (such as colour cameras for television or layers of emulsion for photography). The wavelength-response of these three detectors should correspond as closely as possible to the sensitivity of each of the three cone systems in the retina; and so one should be triggered primarily by blue light, one by green and one by red, together, of course, with a weaker response to light of neighbouring wavelengths. The response of each detector is processed (e.g. electronically in television, and chemically in photography) and coupled to some means of producing light of composition such that it provides an acceptable match of the original. Ideally, this light (which is often a mixture of blue, green and red) should produce exactly the same effect on the retina as does the light from the original scene, although its composition will probably be quite different.

Mixtures of coloured lights can be obtained in a variety of ways. Three projectors fitted with filters of different colours may be used to produce overlapping patches of light on a screen. Maxwell's famous image of a tartan ribbon was obtained in this way by passing three coloured beams through black and white photographic positive transparencies and superimposing them on a screen (but cf. page 177). But multiple projection is suitable only for the large audiences of cinema or lecture hall. Other mixing techniques are needed for books, photographs and television.

Methods of mixing light in order to give specified colours are often classified as either additive or subtractive. As the name suggests, additive methods may involve mixing lights of different wavelength, as when we superimpose beams of light of several colours. More often, however, additive methods involve the mixing, not of different types of light itself, but of the stimuli produced by light of different wavelengths. We can do this if the wavelength of light reaching the eye from one point changes more rapidly than about once every thirtieth of a second. Our visual system can then register only the average of the stimuli, and cannot pick out the components.

We have seen (page 137) that the colours on a painted top become mixed as the top spins. A similar effect is obtained from mosaic patterns of fine dots. When different colours are very close together they cannot be resolved by the eye, and what we perceive is a mixture of the separate stimuli.

Thus a closely packed pattern of red and green dots looks yellow; and, naturally, the greater the distance from which the mosaic is viewed, the more difficulty we have in resolving the separate colours. Mosaic colour mixing of this type was used in early colour photography. It is still used in the printing trade and now forms the basis of colour television, where the phosphors (page 57) providing blue, green and red light are arranged either as a fine mosaic of dots or in a series of closely packed, thin stripes.

Colour photography, like many types of colour printing, is based on subtractive mixing. The original source of light is white (and therefore a mixture of wavelengths). By passing the light through layers of dyed gelatine, or printing ink, light of some wavelength band is subtracted from white light, and so we perceive colour.

We shall discuss the three main media for colour reproduction separately as each has its own problems.

Colour Television

When a scene is scanned by a television camera, the light from the subject is used to activate a transmitter which broadcasts visual information as radio waves. These waves are detected by the receiver, in which they induce small electric currents. The currents are amplified and used to stimulate the light-emitting phosphors in the television tube into producing the picture on the screen.

At any split second, we obtain information about only one tiny area of the scene. In black and white television, the information concerns only brightness; further information is obviously needed if coloured pictures are to be received. But both types of television set work on the same basic principles. The camera scans the scene and obtains information about each of between 100,000 and 500,000 points in it. This information is transmitted, received and converted into a stream of electrons emitted by a gun which moves across and up the screen in exactly the same zig-zag way that the camera scanner moves across the original scene. At any one moment the stream of electrons hits one point on the back of the television tube and makes it glow. The gun moves so rapidly that the eye averages the effect of the light emitted from all points on the screen. The electron beam impinges on every point about twenty-five times a second, which is too fast for the eye to register the dark periods between the flashes. Our eye averages these flashes, and so we see the picture as complete, just as we see any movement within it as smooth.* A black-and-white screen contains two types of phosphors which glow when electrons strike it. One substance gives out

* Given the number of patches per frame, this means that the beam covers between 100,000 and 500,000 patches per second.

blue light, and the other yellow. They are combined in a proportion so that the mixed light emitted is bluish white.

Current colour television cameras contain three light-sensitive elements, and colour television tubes contain three phosphors which emit light of different colours. So two obvious problems are the choice of those colours to which the camera should respond, and the choice of those which are emitted from the tube. A third problem is electronic, and peculiar to present-day television: that of transmitting a signal which can be received by both colour and black-and-white sets to give equally satisfactory pictures.

The selection of colours to which the three parts of the camera are sensitive is dictated by the response of the three cone systems in the retina; by use of suitable filters, together with electronic feedback, the television camera duplicates the sensitivity of the cones as closely as possible.

Though the choice of phosphors is, of course, limited by the emission colours which are so far available, it is useful to consider which colours, if available, would be best to use. We know that any mixture of two lights may be represented on the chromaticity diagram (Figure 76) as a point on the line joining the points for the two original lights. So the point representing any mixture of three lights will lie within the triangle formed by joining the points for the three lights. No colour represented by any point outside the triangle can be produced by these three primary colours. Hence, in order to be able to produce as wide a range of colours as possible, the triangle should cover the greatest possible area of the chromaticity diagram. But, as the area enclosed by all colours is not itself triangular but tongue-shaped, some colours must inevitably be excluded. We might think that it would be best to choose the extreme visible wavelengths such as 450 nm violet and 770 nm red, together with a colour near the apex, such as 515 nm yellow; only the highly saturated edge colours would then be excluded. However, different primary colours (marked as open dots on Figure 76) have been recommended, for two main reasons. We know (page 163) that, at the extreme visible wavelengths, a sensation of reasonable brightness is produced only by light of very high intensity; and so there would be technical (and economic) problems in producing light of sufficiently high intensity to prevent the blues, reds and violets from looking too dull. If yellow were used as a primary colour, the colours in the right-hand edge of the chromaticity diagram would be lost, and those in the gold-orange-red range, as well as those in the green-blue-violet region, would all be somewhat unsaturated. But saturated colours in the neighbourhood of orange occur quite frequently in real life, while greens, blues and violets are usually diluted with white light. So, to represent the types of scene we usually see, it has been recommended that colour television should use as primary colours, the blue, green and red shown in the diagram. In practice, the blue and red

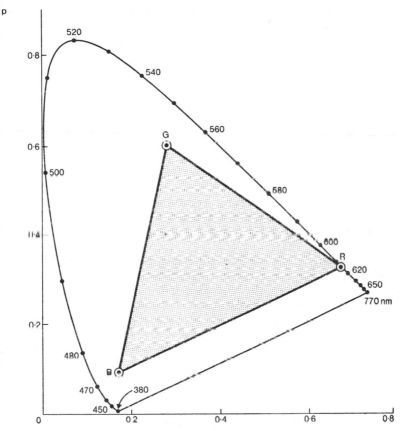

Figure 76. Colour television *The shaded triangle shows colours obtainable by use of the three common phosphors R (europium yttrium vanadate), G (zinc cadmium sulphide) and B (zinc sulphide) (cf. Maxwell's triangle, Figure 66, page 153).*

phosphors approach the recommended points quite closely, although the light emitted by the green phosphor is much less saturated.

Some of the most interesting problems connected with television transmission arise because the same radio signal is used to produce the pictures on both colour and black-and-white screens. The electronic solutions to this problem vary in different countries, but all are based on the fact that the three visible wavelength bands to which the camera responds are related in a rather complex way to the radio signals which are transmitted. In current systems, the brightness of the three separately coloured bands are combined to give the main radio signal which determines the 'luminance' of the picture, i.e. its brightness, whether in colour or black and white. Signals relaying information about colour are obtained by other types of combination of the responses of the camera to the three visual wavebands. In some systems, the information about colour is coded in terms of dominant hue and saturation. Others use blue (of zero brightness) and red (also of zero brightness). When the signals giving colour information are transmitted, they are superimposed on the main, luminance signal in such a way as to have no deleterious effect on the pictures on black-and-white screens. They can, however, be recombined with the luminance signal to stimulate the three differently coloured sets of phosphors. In those systems which code colour information as saturation and hue, these two variables are transmitted simultaneously. The colour signal takes the form of small oscillations above and below the smoother, luminance signal. The height of these oscillations gives the saturation, and the spacing between them gives the hue. Such systems include NTSC, devised in the United States in 1948, and PAL, which was developed from it in Germany, and is now also used in Britain.

Alternatively, only one piece of information about colour may be transmitted at any one time. The French SECAM system transmits, alternately, signals for luminance and blue component, and for luminance and red component. Since the signal for green is obtained by subtracting the blue and red signals from the luminance, the system must contain a 'memory' for storing the value of, say, the blue signal while the red one is being transmitted. The previous colour signal is memorized electronically for 64 millionths of a second.

When the radio signals are received by the television set, they are processed in such a way as to produce the three streams of electrons which stimulate the phosphors to emit blue, green or red light. In older television sets, the phosphors are arranged in a regular mosaic pattern of dots, with the components so close together that the separate colours cannot be resolved by the eye. The dots are covered by a masking screen perforated with 200,000 holes; and the moveable electron guns are mounted at different angles so that the electrons from the 'red' gun can stimulate only the red-emitting phosphors, and so on (see Figure 77). Naturally, such intricate scanning requires careful alignment, and the shielding of each set of dots from stray electrons. In more recent television tubes, the three sets of phosphors are arranged in a series of vertical stripes, covered by a metal grille to ensure that each phosphor is activated only by electrons generated by the appropriate signal. This design is more satisfactory since the problems of aligning the moving electric guns with a striped screen and a grille are much less complex than those for accurate activation of a mosaic screen. The circuits are designed so that response of the phosphor is as closely

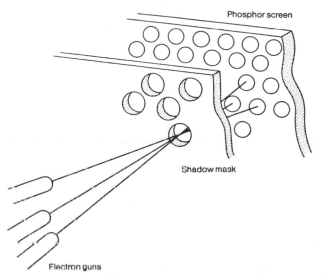

Figure 77. Phosphor dots *The use of a mask to ensure that the electron beams reach their intended targets. (Reproduced, with permission, from R. Osborne,* Lights and Pigments, *John Murray, London, 1980.)*

related as possible to the response of the appropriate component of the television camera. But at the present state of art, a change in colour of the original does not correspond to an identical change in the colour of the television picture. This does not necessarily matter, as the audience does not know the particular colour of a dress, or even exact greenness of the grass. However, the viewer is much less tolerant about the colours of faces. Flesh tints should be 'natural'; neither florid nor anaemic. Many colour television sets have manual controls so that the viewer can adjust the hue and saturation of the colours (as well as the brightness of the picture as a whole). Naturally, most viewers set the controls to give faces of acceptable colour. Some sets have a device for adjusting saturation and hue automatically in order to prevent the colours of faces from deviating from some pre-set value even if, say, the lighting changes or an outdoor camera is exchanged for an indoor one.

Magnetic recordings of television signals are much used both for long-term storage and for the almost immediate recall of the sports broadcaster's 'action replay'. The main problem in videorecording lies in the sheer bulk of information which must be coded – several hundredfold the amount involved in audiorecording. For videorecording in colour, one channel is used for each of the three primaries. In recent years the technology and selling of videorecorders has made immense progress and shows no sign of abating.

Colour Photography

Colour photography, whether for the cinema, for still transparencies or for prints, is based on the subtraction of light of a number of wavelengths from white light. A transparent layer of gelatine, stained with dyes of various colours, acts as a filter, and is placed between the source of white light and some white material from which the coloured pattern is reflected towards the observer. In the projection of moving film or a still slide, the coloured transparency is placed near to the projection lamp and the viewer sees a large image, thrown back from a screen some distance away. In a photographic print, on the other hand, the coloured gelatine is mounted directly on to a reflective base of white paper. The diffuse light which falls on the print and is scattered back to the observer has passed twice through the dyed surface of the print.

Colour photography is nearly as old as black-and-white photography; and both make use of the same method for monitoring light intensity. The manifestation of the colours is a chemical embellishment after the main, photochemical change has taken place, and, for this stage, a large number of processes have been used. Most present-day photography is based on the fact that certain silver salts, after exposure to light and treatment with a chemical *developer*, are changed into grains of metallic silver, the density of the grains being an indication of the intensity of the original light. When any unchanged silver salts are removed by a chemical *fixer*, the pattern of silver grains is now permanent and insensitive to subsequent exposure to light.

Black-and-white photographic prints are obtained in two stages. First a film is exposed and processed, to obtain a partly transparent negative on which pale areas represent dark objects in the original scene and vice versa. A positive print can be obtained by shining a beam of light through the negative on to white paper covered with a light-sensitive layer similar to that used in the film. Light of high intensity comes through the pale areas in the negative, and after development, these give dark patches on the print; as these dark areas correspond to the dark parts on the original, the print is a positive representation of the scene.

Colour photography, too, depends on the formation of silver grains from silver salts which have been exposed to light. The techniques for recording the colour are based on methods devised to ensure that, first, a particular part of the photographic emulsion produces silver grains only if it has been exposed to light in a particular wavelength band; and that, secondly, in the same part of the emulsion, a grain of silver (or an exposed grain of silver salt) can in some way generate, and anchor, a dye of a certain colour. So the photographic material can be persuaded to register the colour of the light which falls on it. If the silver grains are removed, together, again, with any unchanged silver salts, we have a permanent record of the coloured scene.

Colour photography, like colour television, records the intensities of light of three different ranges of wavelength which are roughly those to which the three types of retinal cone cells respond. A direct way of recording colours photographically would be to take three separate photographs (instantaneously and from effectively the same position) through coloured glass filters which are transparent only to blue, green or red light. Then a patch of blue sky would be represented by an area of silver grains in the film exposed through blue glass, but by clear patches in the other two pieces of film. If the silver grains in each layer were replaced with suitable dyes, a complete record of the coloured scene could be obtained by superimposing the three pieces of film. All early colour photography involved the recording of the intensity of three different colours on separate pieces of film,* and such methods are still used in some professional work. But today, the vast majority of colour photography uses film and printing paper in which these three different layers of emulsion, each sensitive to one of the three colour ranges, are already superimposed.

The detailed technique depends on which type of colour photograph we need: a still colour transparency, a motion picture, a colour negative, from which to obtain a colour print, or a xerographic copy. Obviously, for normal use, all the final pictures must be colour positives; a blue sky must be represented by a blue area on the photograph.

We shall illustrate (see Figure 78) the use of a three-layer colour-sensitive emulsion by outlining the steps by which we can obtain, and print, a colour negative. The first layer of film through which light passes responds to light of low wavelength; to blue, and to mixtures (such as white, turquoise and purple) of blue with other colours. The second layer responds to green (and hence also to white, yellow and turquoise); while the third is sensitive to red (and to white, yellow and purple). During processing, the exposed areas are dyed by a substance of colour complementary to that of the light to which the layer responds. So parts of the top, blue-sensitive layer become

---

* The late Dr E. J. Bowen told me that, as pointed out in *Notes and Records of the Royal Society*, 1980, 35, 86, Maxwell's demonstration of colour photography of a tartan ribbon should not have worked since the photographic emulsions then available were not sensitive to red. One hundred years later, some Kodak workers repeated Maxwell's procedure exactly. They found that the red in the ribbon reflected ultra-violet light, which passed through the red filter and affected the film. So it was the ultra-violet which acted as the third primary 'colour', and Maxwell had been lucky.

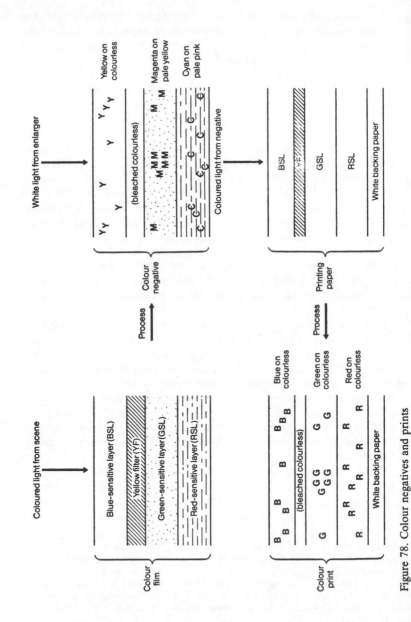

Figure 78. Colour negatives and prints

yellow, of a depth corresponding to the intensity of the blue component of the original light. The exposed areas of the middle layer become magenta when exposed to green light, and those of the bottom respond to red, giving turquoise or, as it is now called, 'cyan'.

The opaque silver grains and the unchanged silver salts are then removed, to leave a film which is clear, and stable to light.

A positive colour print can be obtained from the negative by an exactly analogous process. White light is passed through the negative layer on to the white printing paper, which is coated with a three-layer photosensitive emulsion much like that used to make the colour negative. After processing, the exposed areas of each layer again acquire the colour which is complementary to that of the light which now falls on them (and so the colour they become is the same as that in the original scene). Envisage the Swedish flag, blue carrying a yellow cross, and the red, white and green Italian tricolour, photographed together against a black wall. Table 4 shows the colours generated in each layer of negative film, and in the layers of positive print made from it. The dyes may be thought of as 'negative colours', so that

Magenta M is (white-minus-green)
Cyan C is (white-minus-red)
Yellow Y is (white-minus-blue).

When white light passes through both magenta and cyan, green and red light are absorbed, and only blue is transmitted.

Table 4.

Subject	Film Layer sensitive to			Colour of light transmitted by film	Print Layer sensitive to			Colour of area in print
	Blue	Green	Red		Blue	Green	Red	
Black wall	clear	clear	clear	White	Y	M	C	Black
Blue ground	Y	clear	clear	Yellow	clear	M	C	Blue
Yellow cross	clear	M	C	Blue	Y	clear	clear	Yellow
Red stripe	clear	clear	C	Cyan	Y	M	clear	Red
White stripe	Y	M	C	None	clear	clear	clear	White
Green stripe	clear	M	clear	Green	Y	clear	C	Green

In practice, there are naturally complications. One is that silver salts are particularly sensitive to light of high energy, i.e. to the blue range; so blue light must be prevented from reaching the layers of emulsion which respond to green and red. This is easily accomplished by incorporating a layer of gelatine, dyed yellow, under the top, blue-sensitive layer. The yellow dye absorbs any remaining blue light, and is itself decolorized at the bleaching stage.

A more difficult problem is that the available dyes are not the exact complementary colours of the ranges of light to which the layers respond. Cyan, for example, absorbs not only red light, but some green and blue as well; and magenta absorbs some blue light. So, in a colour negative, cyan areas are contaminated with pink and magenta ones with yellow. In all but highly specialist work, these defects can be largely put right during development by applying a pale yellow masking dye evenly across the green-sensitive layer and a pale pink one evenly across the red-sensitive one. As a little blue and green light is absorbed both by dense areas of say, cyan, and by the clear areas, the balance of colours within the layer is now much improved. It is this combination of the yellow and pink masking dyes which gives colour negatives their characteristic orange cast. The colour may be adjusted further during the printing stage by using filters to make small changes in the composition of the light passing through the negative.

The non-complementary character of radiation and dye is not always a disadvantage; it has been exploited to devise films which are sensitive to infra-red radiation as well as, or even instead of, to visible light. So areas which emit infra-red light are registered as coloured on the processed film. In one such type, tree foliage, for example, appears magenta in the final picture. Most ordinary colour film is also sensitive to invisible radiation, in that it responds to ultra-violet light in much the same way as it does to high-energy visible light, giving a blueish tinge to bright areas such as snow or sunlit sand. This blue cast can be avoided by use of a lens filter which absorbs ultra-violet light but is transparent to visible light.

Much colour photography does not involve an intermediate negative stage. Film for projection, whether of movies or slides, produces a positive image, in roughly the same colours as the original scene. The same is true of Polaroid and other instant prints and of coloured xerographic photocopies. Coloured prints, and duplicate transparencies, may also be obtained from positive movies and transparencies either with or without an intermediate negative.

For colour transparencies and instant prints, a three-layer film is again used, but in each layer the dye of colour complementary to that of the absorbed light is anchored in the unexposed areas, leaving the exposed ones clear. So the top, blue-sensitive layer will be clear in regions exposed to blue, cyan, magenta and white light, but yellow in regions corresponding to black, green, red and yellow, just as in a positive print obtained from a negative (see Table 4). Similarly, the middle layer will show magenta unless it has been exposed to light with some green component. And the bottom layer, bleached by red, becomes cyan in unexposed areas.

As positive colour transparencies are formed in a single stage, there is no possibility of correcting the colour balance in the course of processing. In order that the colours be as 'true' as possible, different types of film are available for use in different lighting. Since photoflood and tungsten lights have a higher proportion of red than does daylight, a transparency of an interior, lit by a tungsten light, will have a reddish cast if it is taken on a film intended for outdoor use. However, if it is essential to use a film which was not designed for that particular lighting, the composition of the light which falls on the film can be adjusted somewhat by using a suitable filter.

The method by which the colours are generated in the proper areas varies with the technique. Processing of transparency film involves, first, developing, and then activating, the exposed silver grains. The film is then re-exposed to light so that the dyes can stain those areas not originally affected.

The Polaroid instant process involves the diffusion of dyes towards the white backing paper; but only those from the unexposed regions are mobile and so able to contribute to the positive image formed on the paper. Dyes from the exposed regions remain trapped in the emulsion, which is peeled away when development is complete.

Xerographic colour photocopying also gives direct positive copies, but without the use of film. As in early colour photography, three exposures are made, using green, blue and red light, but in Xerography, the effect of exposure is to remove electrostatic charge, previously applied to the paper. The unexposed areas retain a charge which attracts a dye of colour complementary to that of the light. After the paper has been exposed, and processed, three times, the three dyes are heat-fused on to the surface to give a permanent copy.

Colour Printing

Most of the colour reproductions around us are neither fleeting images on the screen of television or cinema, nor yet the relatively permanent photographic prints or slides, whether unique pictures or run off in small numbers. The vast majority are produced neither electrically nor chemically, but are printed mechanically by the thousand: as reproductions of advertising posters, designs for postage stamps, cereal packets and beer labels, copies of photographs for book illustrations, for postcards for museums, art galleries and sea-side stalls, for the coloured supplements to newspapers, and for the myriad other pieces of coloured printed matter which we handle daily, frequently with minimal awareness. It might seem that these mass-produced reproductions, often of humble originals, need less skill than that required of the photographic craftsman producing a single, high-quality enlargement in his darkroom. But the business of the printer is usually to make copies of a design, painting or photograph; as later comparison of the printed matter with the original is often possible, the reproduction must be as faithful as possible. The photographer works under no such constraint: except under the strictly reproducible conditions of the copying studio, his subject has gone the moment the shutter closes. So it is never possible to make a rigorous comparison between a photograph and the original scene.

In some ways colour prints which are produced mechanically are similar to those produced photographically. In both types, yellow, magenta and cyan layers of transparent material are all superimposed on a white background. Light passes through the coloured covering and is returned through it again on its way to the observer; and the printing inks, like the photographic dyes, act by subtracting blue, green and red light from the white source. We can see something of the bare bones of the process if we can find some of last summer's postcards or sales posters still on display. The magenta dye is usually the first to fade in the sun, leaving layers of cyan and yellow to provide a picture in shades of turquoise to apple green. Then the yellow is bleached, to give a monochrome in different intensities of cyan.

Obviously, for reproduction of the widest possible range of colour, it must be possible to vary the relative amounts of the three inks in any area of print. In colour photography, this is achieved by varying the concentration of each dye at a particular point (which usually means varying the number of dye molecules for a given area of a layer of fixed thickness). In mechanical colour printing, the inks are applied pure, and at a constant thickness; variation in the concentration of the colour is achieved by not applying the ink solidly over a particular area, but by using a pattern of dots which can be varied in size. Since the dots are too small, and too closely packed, to be resolved by eye, variation in their size gives an effect of shading. These dots form the basis of all half-tone processes, including the reproduction of black-and-white photographs, as we can see by looking at a newspaper photograph through a magnifying glass. The components of blue, green and red light which reach the observer from the original are usually obtained photographically, by means of filters. In addition, a black-and-white positive photograph is obtained, through a yellow filter, which absorbs some of the light in the blue range, to which photographic emulsion is particularly sensitive.*

Each of the three colour separations, together with the positive, black separation, can be converted into a dot pattern with the aid of a very fine screen; and the dot pattern transferred to that part of the printing press which carries the ink. The ink may be applied from raised dots (as in letterpress processes); from dots which are level with the background but differentiated from it chemically (in lithographic printing, the dots are treated with a greasy substance which holds the ink, while the background is soaked with water, which repels the ink); or from sunken dots, which, in gravure processes, act as minute reservoirs of ink.

The analysis of three-colour half-tone prints is extremely difficult. If the dots are separate from each other, the eye integrates the individual colours of the mosaic, and we have additive mixing as in colour television. If, however, inks of two colours overlap, we have subtractive mixing, as in colour photography; and there is the further complication that printing inks are less transparent than photographic dyes and may scatter some of the light which falls on them. They may also be slightly shiny, and reflect some white light from the surface. Since the inks differ from each other in opacity and shininess, the exact colour of the final print depends on the order in which the inks are applied. The thickness of the ink naturally affects the colour, as does the type of process used. Additive mixing plays only a small part in gravure printing, but is much more important in letterpress and lithography. Printing inks often absorb more 'unwanted' colours than photographic dyes, so that magenta ink may, for example, absorb more yellow light than its photographic counterpart. Masking is therefore even more important and many types of electronic monitoring have been used to prepare masks which give more faithful colour reproductions.

Although it is theoretically possible to print black areas by superimposing yellow, magenta and cyan inks, better blacks and more detailed shadow areas may be obtained by using black ink in addition to the three colours. If colours which are under a predominantly black area are removed or weakened, an appreciable quantity of expensive coloured ink can be saved, and the cost of the process reduced. Various suggestions have been made for increasing the range of attainable colours by using processes based

* Coloured filters are also used in black-and-white photography to counteract the effect of the high sensitivity of the film to light of short wavelength. Since the film responds almost as much to blue light as to white, it reduces the contrast between sky and cloud; but if a yellow filter is used, some blue light is absorbed, and so blues are darkened relative to white (and to other colours). Clouds are now rendered white against a darker grey sky. A starker contrast is obtained with an orange filter, while a red filter gives a moonlit effect by dramatically darkening all colours except white and red.

on other four-colour processes, and even on five colours, of which one is black.

When printing with two or more half-tone colours, it is important to avoid patterns of dots which are almost, but not quite, superimposed, because such an arrangement would produce a moiré effect, like watered silk. The dots of different colours are therefore printed so that their ranks are at different angles to the bottom of the page. Black, which is the most prominent of the inks, runs diagonally across the page because, for some reason we do not understand, the eye is least sensitive to dots across the diagonals. (If you look, straight, at a black-and-white photograph in a newspaper, you will see no dots; but if you turn the paper through 45°, so that you have one corner of the picture towards you, it is often quite easy to see the lines of dots.) For yellow, which is the least prominent ink, the lines of dots run parallel to the edges of the page. Other colours are set at intermediate angles.

So half-tone printing in three colours and black is extremely complicated, the total effect depending on the process, the type of ink, its thickness, the angles at which the dots run, the order in which the inks are applied, and methods used for colour correction. But not all three-colour, or four-colour, printing is half-tone. Occasionally, simple designs are printed, not as dots but as large patches of solid colour. Only eight differently coloured areas can be produced:

white	–	no ink
yellow	–	one ink only
magenta	–	one ink only
cyan	–	one ink only
red	–	yellow and magenta
green	–	cyan and yellow
blue	–	cyan and magenta
black	–	all three inks; or black

For this type of work the colour separations are usually prepared by the artist rather than by photography. The earliest coloured illustrations printed in Europe date from the late fifteenth century, not long after the invention of the printing press. The colours were applied without overlap, not from a block, but by stencil.

This chapter has been limited to the discussion of ways in which we try to make as good an imitation as possible of the colours of some original scene, picture or object. Man's role here is purely technological. But in subsequent chapters we shall look at some of the more creative ways in which we use colour.

17

ADDED COLOUR

Although much of the view from the window is man-made, most of the colours are natural, or nearly so. For sky, plants, stonework, the colours are outside man's control. The colour of brickwork depends, of course, on the clays from which the bricks were made and on the temperature at which they were baked; and it is only for the colour of the exterior paintwork that man can take sole responsibility.

Indoors, however, the situation is quite different. Maybe you are sitting at a table of natural pinewood, laid with rush mats, metal cutlery and colourless glasses. On a shelf are some shells and an earthenware vase of dried flowers. You are, perhaps, wearing an Arran pullover. To none of these objects has any colouring matter been added; but these are exceptions. Walls are painted or papered, mugs glazed and curtains dyed; and there are pictures on the walls. Women wear make-up and cakes have coloured icing.

The great majority of objects produced in factory, workshop or studio have been coloured artificially, sometimes only on the surface (since colouring materials are often expensive), but sometimes all through. The main function of a surface layer is often to act as a sealant against water seepage. Untreated wood holds water in which rot-fungi thrive, and liquids leak slowly through unglazed pottery. But mankind learnt early that pottery glazes need not be purely protective: they could simultaneously be used, with infinite variety, for decoration. For many jobs, paints are more durable than, for example, varnishes, and with the use of paint comes a choice of colours.

For many coloured objects, however, the colour is not a mere improvement to a protective outer layer which will in any case be added. Cloth is no more durable for being dyed (indeed, the converse is true), but it is certainly the more enjoyable. As we shall see, there may be a number of reasons for wearing make-up, for painting pictures, for wearing clothes of a particular colour and for using coloured lights in theatres, in discos and at road junctions; but in none of these cases is durability of prime importance.

In this chapter, we shall look briefly at the ways in which we can use dyes and pigments to add colour to an object, whatever our reasons for doing so. A dye is a highly coloured substance which dissolves in the medium in which it is applied, so that the individual dye molecules are evenly distributed throughout the dyeing solution or in a transparent plastic. A pigment, on the other hand, does not dissolve in the medium, but consists of particles which are much larger than individual molecules. Pigments are applied as a slurry, or suspension, of small granules in some liquid, such as water, linseed oil or acrylic resin.

As with the reproduction of colour, many of the more interesting problems facing the colourist are particular to the technique which is being used. But there remains the challenge of matching the colours of two objects which are made of different materials; the avoidance of metamerism; and the problems of obtaining an acceptable match between surfaces of different reflectivity.

Paints

Most paints, whether for architectural, technological or artistic use, consist of fine particles of pigment suspended in a liquid, or ground with some liquid medium to form a paste. A thin layer of paint spread over the surface eventually sets to a hard film by the evaporation of some or all of the liquid, or by aerial oxidation or other chemical action. The colour of the paint depends, of course, on which components of white light are scattered back to the observer by the particles of pigment.

The size of the particles is important. If they are much smaller than the wavelength of light, they scatter blue light preferentially; all pigments ground infinitely small would look sky blue (see page 65). Larger particles scatter, unselectively, light of all wavelengths which is not absorbed; and the more particles there are, the better. The best balance between good, non-selective scattering and a large number of particles is usually achieved if the diameter of the particle is slightly less than the wavelength of visible light.

The power of a pigment to cover a dark surface depends on the material which surrounds the particles as well as on the particles themselves. We have seen that, when light passes from one material to another, it is bent. The more sharply the light is bent on passing from surroundings to pigment, the better is the so-called hiding power of the paint. As light is bent more sharply when passing into a pigment particle from air than from water, the colours in a dry watercolour painting are less transparent than they were when wet. On the other hand, as most oil paints dry the sharpness of the angle decreases and the film becomes less opaque as it sets. When light passes from any medium into a particle of titanium white, which is used in 'Brilliant White' paint, it is bent very sharply indeed, and so paints made from this pigment have very high covering power.

Not all coloured, highly opaque substances are suitable for use as pigments: and those that can be used are often better in some media than others. The pigment can best be applied as a paint if it can be properly wetted by the medium; and as a solid if it can be made into a water-bound or oil-bound paste to form a pastel, or incorporated into wax, or clay, to form a crayon. The pigment should not weaken the dry paint film so that it cracks, and it should, of course, keep its colour for as long as is required of it. The colour of paint for exterior woodwork of houses should stay more or less the same for a few years; and, ideally, the colours of a work of art should stay exactly the same for centuries. The main enemies of colour stability are sunlight and atmospheric pollution; but colours can deteriorate also by chemical reaction between two pigments if they are mixed together or are adjacent to each other.

Pigments may be animal, vegetable, mineral or synthetic; and while some man-made ones are closely related to natural pigments, others are purely laboratory products. The first known pigments, from the neolithic cave paintings, were natural minerals. The ochres and reds, and the occasional green, owe their colours to different forms of iron. Chalk served as white and charcoal from the hearth provided black. Later, between 2000 and 1000 BC, other natural minerals were added:

azurite	blue	} forms of basic copper carbonate
malachite	green	
cinnabar	scarlet	a form of mercury sulphide
orpiment	yellow	} forms of arsenic sulphide
realgar	orange	

The second millennium BC also saw the manufacture of the first two synthetic pigments: white lead, made by treating sheets of lead with vinegar; and blue frit, a pulverized glass, coloured with copper ore. Although blue frit was much used in ancient Egypt and in Minoan and classical antiquity, its use lapsed between AD 200 and 700. White lead, however, remained the supreme white pigment until the manufacture of 'Chinese White' from zinc in the 1830s.

Classical antiquity saw the introduction of the first organic pigments. Verdigris ('Green of Greece') was synthesized for use as a pigment by exposing sheets of copper to the vapour given off by fermenting grape-skins. Some, like indigo from the woad plant, and Tyrian purple from murex, a type of whelk, were also used as dyes (see page 191). The writing inks of antiquity might be of animal, vegetable or mineral origin: sepia from the cuttlefish; mixtures of iron salts with gall, or with oak bark; and various forms of carbon in different media.

These pigments had to suffice for many centuries, and it was not until after A D 1200 that the artist's palette was enriched by synthetic vermilion (which alchemists made by combining sulphur and mercury); a yellow waste product from the glass industry; and natural ultramarine, made by grinding the blue particles in lapis lazuli. 'Lake' pigments seem to have been first used in this period; a soluble dye, such as the red extract of madder root, is converted to an insoluble pigment by adsorbing it on to some insoluble, but almost transparent, base. Bases of this type may also be used without adsorbed dyes, as 'extenders' to improve the mechanical properties of a paint. Saffron, from crocus flowers, was used in illuminated manuscripts, and gamboge, a golden pigment from a plant resin, was introduced from the East. The organic pigments which were then known tend to fade badly in sunlight, but this is obviously much less serious for book illumination than for paintings which would be constantly exposed to daylight.

Cochineal, from an insect, was introduced after the discovery of Central America, and used as carmine lake. As a water colour, it too faded. But it was more stable as an oil colour; and the related crimson and purple lakes were much used as coach colours.

There was another hiatus before the next new colours, which were all synthetic inorganic pigments: prussian blue (1704), cobalt blue (1802), emerald green (1814), chrome yellow (1820), synthetic ultramarine (1828) and zinc white (1830s). Mixtures of chrome yellow and prussian blue were, and still are, used to make paints of a variety of different greens. The yellow particles contain lead, and in an oily medium the lighter, blue particles tend to collect near the top, making the unstirred paint look bluer than the stirred mixture. The two pigments deteriorate differently, depending on conditions. On exterior work, the yellow tends to fade the more, and so outside doors, painted green, often become more blue as they age.

A knowledge of the history of pigments is essential for the reliable attribution of a work to a particular artist and for the detection (or perpetration) of art forgery. It emerged in a court case that a *Merry Cavalier*, thought to be the work of Franz Hals, was painted using pigments which included zinc white and synthetic ultramarine, against a large background of cobalt blue. But Hals died in 1666, before these were in use.

From the 1850s onwards, a new type of synthetic pigment was introduced; a number of synthetic organic substances, related to madder, were made from coal tar. They included some mauves and purples, brighter than those which could previously be made only by the mixing of reds and blues. Unfortunately, many synthetic organic colours are as fugitive as their naturally occurring counterparts; and they were sold to artists before they had been sufficiently tested. Gone were the days when painters, or their apprentices, prepared their own pigments using recipes which had been tested by centuries of use. There was considerable friction between artists and suppliers in the latter half of the last century. Holman Hunt, in particular, conducted a long campaign against the colourmen. It took about fifty years before pigments of acceptable stability could be prepared from coal tar, but several reliable colours in the orange, red and purple range have now been made.

Another type of synthetic pigment was copper phthalacyanine, or monastral blue, first made in 1936. It is related to chlorophyll; and it is quite stable. Within a year or so, it was adopted as the signature colour of the newly hatched Pelican Books.

Metal flakes and powders may also be used as pigments. A suspension of powdered gold has been used as a paint since antiquity. With metals less noble, the dried medium or coating must protect the metal from corrosion. Not surprisingly, silver paint tends to blacken if inadequately varnished, and has now been replaced by aluminium. Copper and bronze powders are also used in paint manufacture.

If paintings are to remain unspoiled over the centuries, it is essential that the artist use not only stable pigments, but also media and varnishes which do not deteriorate with age. Surface dirt is only a small problem compared with a discoloured varnish. Some yellowing is not too deleterious with colours in the red, yellow, brown range, or even with green; but it kills a good clear blue. Early paintings which incorporated natural ultramarine were particularly vulnerable because the pigment was ground only coarsely, in order for the deep colour to be preserved and not diluted by too much scattered white light. When the pigment particles are large, there are sizeable, varnish-filled spaces between them and the effect of any yellowing is much aggravated. It is even more important that the varnish remains unbroken. The bad deterioration of Reynolds's portraits, often of now-pallid faces covered with a network of cracks, is more attributable to inferior varnish than to his use of fugitive pigments for flesh tones.

The use of paint is not restricted to buildings, objects and works of art. People are painted, too, be it with tribal warpaint or sophisticated modern cosmetics. Body paint has developed from pastes and powders of coloured muds, through pastels of rarer minerals to the present-day paints, crayons and pastels meticulously manufactured from non-toxic synthetic dyes, for use as nail varnish, mascara, lipstick, eye-shadow, rouge and face-powder. Although these may contain various non-toxic synthetic dyes, iron oxide is still a major constituent of the colouring used in rouge and face powder. The 'frosted' appearance of some lipsticks and nail varnishes may be provided by adding flakes, sometimes of aluminium, sometimes of natural substances such as mica or mother-of-pearl. Today, the function of cosmetics may be purely social, but in ancient Egypt they also served to reduce fly-borne diseases of the eye. Early jars containing green malachite (for the lower eyelid) and black kohl (for the upper one) bore medical inscriptions; but later these colours were replaced by a green vegetable dye and charcoal, and the inscriptions became purely cosmetic.

The cosmetic arts are not restricted to the use of superficial removable colours. From earliest times, tattooists have decorated the human body by implanting insoluble particles under the skin to produce designs which can only be removed with the greatest difficulty. With the exception of the ubiquitous synthetic blues and greens (see page 188), the pigments are familiar paint-box colours, such as carbon, iron oxide, carmine, scarlet lake and zinc white. Others, such as cadmium yellow and cobalt blue, are thought to sensitize the skin; and with vermilion (mercury sulphide), a sensitive reaction can result after delays of up to forty years. The range of colours employed by tattooists varies with the culture, as does the purpose of the tattoo, and the design. The elegant, precisely symmetrical spirals which adorned the faces of Maori warriors were often in black alone, while Englishmen have been known to have large fox-hunting scenes implanted on their backs in full, if somewhat crude, colour. Japanese tattoos may cover the entire body, often with some abstract dragon designs executed in such subtle colours that the effect is reminiscent of exquisite lacquerwork, or inlays of exotic woods and mother-of-pearl.

Dyes

Dyes and stains differ from paints in that they penetrate much more deeply into the material they colour instead of adding a layer of pigment to the surface. Stains, which are used mainly for wood and leather, penetrate the material for a short distance, whereas true dyes permeate the material fully, giving it the same colour all through. Although dyes were probably used first on leather, we shall confine most of our present discussion to the dyeing of textiles.

All textiles are made from fibres: animal, vegetable or man-made; and the art of dyeing consists of persuading a coloured substance to become intimately associated with the fibre, in such a way that the colour is evenly distributed on the cloth, does not wash out, fade in sunlight or have any ill-effects (such as making the cloth harsh, non-absorbent or prone to rot). The colour may be added to the fibre at various stages in the manufacture: the object which is dyed may be the final garment or curtain, a length of cloth, a hank of thread or short unspun natural fibres; or dye may be dissolved in a mixture used to extrude synthetic fibres. In some ways, textile fibres are rather like matted seaweed, dried up on a beach. They may be strap-like or branched; and they may be very tangled, or laid one on the other in a moderately orderly fashion. When the lump of seaweed is put into water, the pieces swell and push each other apart, enlarging the spaces between them.

Most textile dyes are organic molecules which are soluble in water. Although, on the molecular scale, they are by no means small, they are very much smaller than even the most finely ground particles of pigment. When textile fibres are soaked in an aqueous solution of dye, molecules of the dyestuff permeate the spaces between the fibres; and as the fibres dry, the water evaporates and the dye remains trapped. A satisfactory dye will remain trapped even when the fibres are immersed in water.

The first dyes, some probably used over 5,000 years ago, were natural organic colours. Both safflower (from a type of thistle) and indigo were used to dye the bindings of Egyptian mummies. Somewhat more recently, around 3,500 years ago, Minoan frescoes at Thira and Knossos show the crocus from which saffron was harvested, and the saffron-gatherer. Another vegetable dye much used in the ancient world was madder. Two important animal dyes were also in use at this time: Tyrian purple, from the whelk-like murex, and kermes scarlet. The kermes insects live on oak and holly and were harvested at night by women who carried lanterns and kept their fingernails long especially for the purpose. The dried insects are said to be on sale in Athens to this day. The discovery of the New World produced new vegetable dyes such as logwood and fustic, and also cochineal, from a beetle similar to kermes.

Many of these natural dyes, except for indigo, would not combine permanently with untreated fibres. The cloth, or yarn, was first impregnated with some simple inorganic substance (usually a metal salt or hydroxide) to which the dye adheres strongly. These so-called mordants, on which the dye can bite, play the same role, and are often the same substances, as the transparent materials on which 'lake pigments' are formed (see page 188). The final colour depends on both the dye and the mordant, and, to a lesser extent, on the fibre.

The colours of non-mordant dyes, such as indigo, could be varied somewhat by first dyeing the cloth to a shade darker than needed and then washing out the dye until the required colour was reached. The cloth was then treated in some way which made the colour more fast. Although these natural dyes of antiquity were the only ones available until the nineteenth century, there was considerable progress in ways of opening up the fibres (by treatment with heat or alkali) to allow entry of both mordant and dye, and in methods of fixing the dye subsequently, often by steaming.

But the major advances, from the mid nineteenth century, followed the synthesis of dyestuffs from coal-tar products. The teething troubles were considerable. Many of the new dyes, like the first coal-tar pigments, were very fugitive; and it was over half a century before they were even as fast as the old vegetable ones. The first synthetic mauve dyed silk so rapidly that it was impossible to obtain an even colour. Nowadays, however, synthetic dyes have largely replaced the natural ones for industrial work; over 7,000 of them are in current use. Details are compiled into the vast 'Colour Index', and the effects of dilution and mixing dyes have been charted in 'colour maps', colourists' grids related both to sections through colour solids and to CIE diagrams. Large numbers of different techniques for applying the dyes have been, and are still being, developed. Dyes may now be added to fibres from a vapour or from organic solvents, rather than aqueous solution, and synthesized during the dyeing process. Sometimes the dye can be bound to the fibre by strong chemical links. The dye may alternatively be added with the finishes, almost as if the fibres were coated with some coloured varnish which had little or no harshening effect on the cloth.

But problems remain. Even modern dyes are not totally fast to washing or to sunlight. Loss of fastness is not, however, always a disadvantage, as is shown by the popularity of faded blue denim, a colour achieved by wear, sun and washing on cloth dyed with synthetic indigo. Sometimes the light energy absorbed when a dye is bleached by the sun breaks down some of the chemical links in the fibre; the faded parts of the cloth then weaken and rot. Clothes containing those dyes which are sensitive to temperature may change colour alarmingly when they are ironed; but since the change is fully reversible, no permanent harm is done. Towels which are dyed in strong colours are often less absorbent than towels of pastel shades; when the cloth is dyed, the molecules of dye take up positions which could otherwise accommodate water molecules.

Some interesting dyeing techniques were evolved to meet the need of the craft of textile printing, which in Europe dates from the twelfth century. Earlier 'coats of many colours' were obtained by weaving, using yarns which had been dyed separately. Several primitive methods for dyeing some parts of the cloth but not others are still in use today. Dye may be kept from certain areas by covering them with wax (as in batik), or, nowadays, with resin; or by tying them firmly with string. Sometimes cloth is first dyed a single colour, and then a white pattern applied by bleaching certain areas, with chemical bleaches, or, in traditional African work, with a concoction of soap and mud. When textiles are to be printed with mordant dyes, the design is applied by treating the cloth with different mordants and then by immersing it in a single dye. Madder, for example, can yield a whole range of pinks, reds, purples and browns in this way, and weld gives a number of colours in the yellow, ochre and brown range. In early mordant printing, most colours were obtained in this way. The exceptions were blue and green, which were added by brushing the non-mordant indigo over white, or yellow, areas.

Textile printing has naturally advanced. The original wood blocks were replaced by copper plates, rollers and stencils, and today screenprinting is widely used. Dyes for printing, as opposed to those merely for dyeing, must be viscous so that they do not spread by capillarity into neighbouring parts of the pattern. One way of achieving this is to give the dyes a dough-like consistency and make up the pattern in the form of a mosaic of these doughs on a printing roller. Any number of colours may be incorporated, but since the dyes become used up, the method is unsuitable for printing very long lengths of cloth. Many different dyes, often of unrelated chemical type, may be used to print a single design. Much cunning is needed to ensure that the dyes are applied by such processes, and in such order, that no stage has a harmful effect on any previous one, and that the finishing process is compatible with all the dyes used.

Textile printers also make use of colourless dyes, known as optical brighteners. These absorb ultra-violet radiation and use some of the energy in vibrating more violently. The rest of the energy is re-emitted as a weakly visible fluorescence, often with a bluish tinge. The earlier use of bluebag (synthetic ultramarine) to whiten the Monday wash worked by removing some red reflection and restoring a colour balance towards white, albeit a white of reduced intensity. Optical brighteners add blue light to compensate for loss caused by yellowing, and so give a much brighter white.

Some of the vagaries of colour perception have been domesticated for economic advantage. The impact of a printed textile design can be totally changed merely by changing the colour of the background, and this may be effected, quite cheaply, by changing only the bottom roller in the printing process. So we can get several apparently quite different prints for barely more than the price of one. The modification of one colour by its neighbours is similarly employed in weaving, where the colour of the same weft yarn seems to change with that of the warp. The pleasing result is that the variety of colours in the weave appears to be much greater than the number of different yarns it in fact contains.

Many other materials may be dyed throughout, or stained on the surface by techniques similar to those used on textiles. Since spaces between the fibres in wood cannot be widened in the same way as in textile fibre, dye cannot penetrate so readily and wood can be coloured only superficially. Leather, too, has only small interstices in its structure, as has live human skin and nails. The ancient Egyptians are thought to have considered it a social necessity to stain their nails, palms and soles with henna, a reddish dye extracted from a conifer. Some metals may be 'dyed', or more precisely, stained. Aluminium, for example, can be treated electrically so that it acquires a firm surface film of its oxide or hydroxide. This coating acts as a mordant, as do metal hydroxides in textile dyeing, and can combine permanently with a wide variety of dyestuffs, to provide a range of brightly coloured metal objects, such as ashtrays.

The colour of hair may be changed by bleaching, or dyeing, or both. Bleaching usually involves treating the hair with hydrogen peroxide, or some other oxidizing agent which converts the melanin (see page 91) to a colourless substance. Hair is dyed by a technique similar to the dyeing of other fibres. It is first treated to expand spaces between the scaly outer coating of the hair and the continuous inner strand, to which the dye molecules are attracted. When the outer coating reverts to its original position, the dye is trapped. Henna was probably the earliest hair dye used, not only for women's hair, but also for the manes of horses and the beard of the Prophet. Later, the bright orange tinge of henna was often darkened with indigo, or lightened with a yellow dye from camomile. A wider range of colours is available today. In addition to a rich variety of natural tones, blue rinses are used to neutralize any unattractive yellowing of basically grey or white hair. Ageing hair is sometimes darkened with a 'colour restorer' which is normally a colourless solution containing lead, silver or copper. In contact with air, it gradually forms a dark coating of the metal oxide or sulphide around each hair. The same process was used in ancient Rome by both men and women who darkened their hair by using lead combs dipped in vinegar.

Glass and Glaze

We have seen that, when we apply colour, there is a wide variation in the intimacy of the relationship between the colouring agent and the object. A layer of paint is only skin deep. Stains penetrate further, but are still superficial. Fabric dyes become entangled with the individual fibres rather than coating the woven or knitted fabric; some dyes even become joined to the fibres by forces as strong as those within the fibres; so these dyes become chemically incorporated in the fibre.

In coloured glass, pottery glazes and vitreous enamel, the colouring is associated even more intimately with the material. The dye is dissolved in the solid glass in much the same way as salt dissolves in water. Solid glass (see page 30) consists of an irregular cagework of silicate groups (each made up of silicon and oxygen atoms). Incorporated into this muddle are ions of various metals such as sodium, potassium and aluminium. Were these the only metal ions present, the glass would be colourless; but the sand from which the glass is made almost always contains a trace of iron, and it is this which gives cheap glass its characteristic pale green colour. Stronger colours can be introduced by adding other metals, or metalloids, usually in the form of the oxide. Blue ('medicine bottle') glass contains cobalt or copper, while 'bottle green' is produced by chromium, iron and manganese. Iron, in the oxidized form familiar from ochre and terracotta earth colours, gives amber glass, while nickel produces purple or brown. Selenium, a metalloid rather than a true metal, gives red, and is used not only to produce red glass, but also to 'mask' the green cast caused by traces of iron. A trace of added selenium converts pale green to a dull straw colour. (Light which has encountered both the iron and the selenium would be whitish, but of reduced intensity; that which had met only the iron would be green; and that which had interacted with only the selenium would be red. The emergent mixture of white with very low-intensity green and red would look very pale yellow.) This final trace of colour is removed by 'blueing' with a trace of cobalt, which does for the glass exactly what bluebag does for slightly yellowed white sheets (see page 193). The slightly grey tinge in the final glass is a little darker, though less obtrusive, than the original pale green from the iron.

Not all coloured glass contains dissolved metal oxides, with the metal ions far apart from each other in the silicate network. Colours in some glass are produced, not by absorption, but by scattering; and so, as we saw in Chapter 4, they depend on the diameter of the particle. The familiar, deep ruby of Venetian glass is produced by suitably sized particles of metallic gold or copper, or of cadmium selenide. Yellow light may be similarly scattered by cadmium sulphide particles.

Coloured vitreous materials have been used since ancient Egyptian times in a wide variety of contexts. Small pieces of glass were used first as beads for jewellery, and as fragments for mosaics; ground blue glass was used as a pigment; and layers of vitreous material formed the glaze on pottery and enamel on metal. Later, storage vessels and drinking glasses were fashioned from glass, and the last millennium has seen the glory of the stained glass window. In all these diverse forms, the colouring matter is fused with the vitreous material, although the details of the methods of producing, and introducing, the colour naturally depend both on the medium used, and the stage of evolution of that particular craft.

In the earliest stained-glass windows, for instance, each piece of glass was of roughly uniform colour, cut from a larger sheet which was coloured by adding a particular metal oxide to fused glass. Faces and other details were painted on to the glass with a paste made largely of powdered rust, and the piece of glass was then heated to a temperature high enough to turn the iron pigment almost black and to incorporate it permanently into the surface, but not high enough for the glass to melt, or to be misshapen. In somewhat later work, a coloured layer was added to one side of a piece of plain glass. Ruby glass was made by 'flashing', that is, by dipping one surface of a piece of plain glass into molten coloured glass. Ruby glass owes its colour to the scattering of light by minute particles of copper, and is so intensely coloured that, if it were of the thickness necessary for a windowpane, it would be opaque. Later still, stained glass of other colours was made by flashing; details could be rendered by abrasing parts of the coloured layer to reveal the colourless supporting pane, in much the same way as bleach is used to give light areas on dyed textiles (see page 192). Colours in the lemon to amber range were also added as a superficial layer, but by 'silver staining' rather than by flashing. The glass was painted with a solution of a silver salt and then heated to incorporate the silver into the surface of the glass. The exact colour depends on the conditions, and, of course, on the colour of the original glass; rich greens were made by staining blue glass.

The colour of glassy materials depends not only on the particular colourant which is fused into the glass, but also on the conditions used and on the composition of the glass itself. When the glass is needed for such large structures as windows, its composition is determined largely by the mechanical properties which are needed. If the vitreous material is needed only for closing the pores of a pot or for decorating the lid of a snuff-box, its strength and durability are less important and more liberties may be taken with its composition. Nor, for such purposes, need it be transparent. The addition of tin and lead oxides, for example, produces an opaque white glass, while a tin and antimony mixture can render a translucent colour opaque.

In the earliest enamels, thin metal strips were fixed to a metal base, and the areas between them filled with powdered glass of various colours. The object was then put into a furnace hot enough to fuse the glass. The composition of the glass could vary from one area to the next. For example, the same black copper oxide produces a turquoise enamel in a glass rich in sodium carbonate but a yellowish-green one if the glass contains red lead. These relatively simple, cloisonné enamels are thought to have been the forerunners of leaded stained-glass windows and also of a variety of more sophisticated enamelling techniques. The metal base can be gouged out to different depths which, when filled with translucent coloured enamel, produce colours of different intensity. The base may be removed from the finished enamel, to give the effect of a stained-glass window. In some modern Japanese work, each area is fired separately and then the surrounding strip removed, so that the various colours touch each other without running. Irregularly shaped objects, such as metal figures, can be enamelled by placing them in a mould, which they just fail to fill. The gap between the object and the mould is packed with powdered enamel, and the whole is fired. Some enamels are painted on to a light, opaque ground, each colour being fired separately to prevent running. The pigments are finely ground enamels, made up into a paste with water and glue. Alternatively, white enamel may be painted on to a dark ground, thickly for light areas and more sparingly for darker ones. Detail may be added by hatching with a needle to produce lines revealing the darker ground.

Glazes for pottery, like enamels, can be opaque, of translucent colour or totally transparent. Colours are again provided by metal oxides, fused into the glass, and an opaque white is again provided by tin. The most primitive way of producing a coloured, glazed pot is to use a single, coloured glaze. But the colours are more often applied either under or over a colourless or very lightly coloured glaze, and may, of course, be painted in very elaborate design. The ubiquitous china blue is achieved with cobalt. For glazes containing oxides of some metals, such as copper, iron and manganese, the colours depend on the atmosphere within the furnace. If there is a generous supply of air, the copper becomes blue, iron gives colours in the yellow to rust range, while manganese is reddish purple. In a so-called 'reducing' atmosphere, deficient of air, copper turns green or 'bluish-red' while iron and manganese become blackish. The red and black designs on Attic vases were produced using clay from a single source but with differing textures which varied its accessibility to air and other furnace gases during the multi-stage firing process. In Classical Greece, as in ancient China, a reducing atmosphere was produced by throwing wet wood on the kiln, and modern potters can copy these conditions in non-electric furnaces which are adjusted so that the internal atmosphere contains unburnt fuel. Porcelain may also be gilded with non-tarnishable metals under the final glaze; the lustre ware produced at Sèvres and Chelsea was gilded by applying finely divided gold, ground with honey. 'Silver' lustres are made using platinum.

Materials of glassy appearance may now be made from polyacrylic resins, and coloured with either translucent dyes or opaque pigments. As in the staining of glass, colouring matter is added to a colourless liquid, which then sets: but, unlike glass, the resin is liquid at room temperature. Since setting is accomplished not by cooling but by adding a catalyst, the colours do not have to withstand high temperatures. The craftsman in acrylic resins is lucky enough to have a wide range of brilliantly coloured synthetic organic dyes, which would, of course, decompose at the temperature needed to fuse glass. Although the artistic use of these resins is as yet at an early stage of evolution, the technique has wide and exciting possibilities.

PART SIX:
USES AND LINKS

Men in a state of nature, uncivilized nations, children, have a great fondness for colours in their utmost brightness, and especially for yellow-red: they are also pleased with the motley.

— GOETHE, *Theory of Colours*

The chief aim of colour should be to serve expression as well as possible.

— MATISSE

It was the poetry of colour which I felt, procreative in its nature, giving birth to a thousand things which the eye cannot see, and distinct from their cause. I did not, however, stop to analyse my feelings – perhaps at that time I could not have done it.

— ALLSTON, on visiting the Louvre, 1803

And among colours there are certain friendships, for some joined to others impart handsomeness and grace to them. When red is next to green or blue, they render each other more handsome and vivid. White not only next to grey or yellow, but next to almost any colour, will add cheerfulness. Dark colours among light ones look handsome, and so light ones look pretty among dark ones.

— ALBERTI (1404–72)

Impressionism

By making up his palette of pure colours only, the Impressionist obtains a much more luminous and intensely coloured result than Delacroix; but he reduces its brilliance by a muddy mixture of pigments, and he limits its harmony by an intermittent and irregular application of the laws governing colour.

Neo-impressionism

By the elimination of all muddy mixtures, by the exclusive use of the optical mixture of pure colours, by a methodical divisionism and a strict observation of the scientific theory of colours, the neo-impressionist ensures a maximum of luminosity, of colour intensity, and of harmony – a result that had never yet been obtained.

— PAUL SIGNAC

My choice of colours does not rest on any scientific theory; it is based on observation, on feeling, on the very nature of each experience. Inspired by certain pages of Delacroix, Signac is preoccupied by complementary colours, and the theoretical knowledge of them will lead him to use a certain tone in a certain place. I, on the other hand, merely try to find a colour that will fit my sensation.

— MATISSE

I should like to paint the portrait of an artist friend, a man who dreams great dreams, who works as the nightingale sings, because it is his nature. He'll be a fair man. I want to put into the picture my appreciation, the love that I have for him. So I paint him as he is, as faithfully as I can, to begin with.

But the picture is not finished yet. To finish it I am now going to be the arbitrary colourist. I exaggerate the fairness of the hair, I come even to orange tones, chromes, and pale lemon yellow.

Beyond the head, instead of painting the ordinary wall of the mean room, I paint infinity, a plain background of the richest, intensest blue that I can contrive, and by this simple combination of the bright head against the rich blue background, I get a mysterious effect, like a star in the depths of an azure sky.

— VAN GOGH, at Arles, 1888

> The road at the top of the rise
> Seems to come to an end
> And take off into the skies.
> So at the distant bend
>
> It seems to go into a wood,
> The place of standing still
> As long the trees have stood.
> But say what Fancy will,
>
> The mineral drops that explode
> To drive my ton of car
> Are limited to the road,
> They deal with near and far,
>
> But have almost nothing to do
> With the absolute flight and rest
> The universal blue
> And local green suggest.

— ROBERT FROST, 'The Middleness of the Road'

> Ah, what shall language do? Ah, where find words
> Tinged with so many colours...

— JAMES THOMSON, 'Spring'

I present a fine case of colored hearing. Perhaps 'hearing' is not quite accurate, since the color sensation seems to be produced by the very act of my orally forming a given letter while I imagine its outline. The long *a* of the English alphabet (and it is this alphabet I have in mind farther on unless otherwise stated) has for me the tint of weathered wood, but a French *a* evokes polished ebony. This black group also includes hard *g* (vulcanized rubber) and *r* (a sooty rag being ripped) ... In the green group, there are alder-leaf *f*, the unripe apple of *p*, and pistachio *t*. Dull green, combined somehow with violet, is the best I can do for *w* ... Finally, among the reds, *b* has the tone called burnt sienna by painters, *m* is a fold of pink flannel, and today I have at last perfectly matched *v* with 'Rose Quartz' in Maerz and Paul's *Dictionary of Color*.

– NABOKOV, Speak, Memory

18

IMPARTING INFORMATION

In this chapter, and the next, we survey the ways in which we use our ability to distinguish between colours, both in the natural world and in our man-made surroundings.

Colour vision must have been a considerable asset to early man in his search for wholesome food; for the development of agriculture, cooking, ceramics and metalwork; for predicting the weather; and for the recognition of his fellows. Today, we use colour to detect mouldy bread, unripe fruit, bruised lettuce, unfresh meat. 'Enough blue sky to make an elephant a pair of trousers' heralds a welcome break between showers. We heat the grill to red heat, and turn it down when the food 'looks done'. We may fail to recognize an acquaintance if, after an absence, she reappears as brunette instead of blonde. We can guess that changes in colour played an important part in primitive technology; and we know that it played a major role in medieval science and technology, as it does today. Despite proliferation of automatic monitoring and control in laboratory and factory, visual observations of colour changes are still important in experimental science and technology, to follow a gradual process, or to observe a sudden change. We have seen how sensitive is the colour generated by some particle to the arrangement of neighbouring particles, as in the use of cobalt compounds to detect water vapour (see page 81). Differences in colour often allow us to recognize the presence of different substances. The coloured rings formed round an ink blot on blotting paper show us that the ink contained a number of dyestuffs which were not only of different colour, but which had been partially separated because they travelled through the paper at different rates. The coloured edges of perspiration stains have a similar origin. This separation of dissolved substances at the edge of a seepage has been developed into a powerful chemical technique, appropriately called 'chromatography' since it originated with observations of coloured bands. The colours are sometimes induced. Microscope slides are stained to reveal different types of cell. Chemists add indicators which change colour with acidity or with a sharp change in the composition of a solution: the first such indicators were vegetable dyes, discussed in Chapter 9. Starch is often added to solutions to indicate, with a deep blue-black, the presence of iodine.

Experienced observers can often detect very small variations in colour, but these, of course, depend on the illumination and on the surroundings as well as on the object being viewed. It has been suggested that hospitals should use uniform lighting, so that any change in a patient's appearance be more easily noticed; and that secondary reflection by strongly coloured surroundings should be avoided. Jaundice can seem to set in rapidly when yellow cubicle curtains are drawn. Tobacco sorters in Bulgaria grade the leaves against a background of very similar colour in order to minimize the effect of simultaneous contrast. Butchers' shops have been known to choose lighting with a generous red component, so that the meat looks appetizingly fresh. If a colour judgement is to be carried out in more than one type of light, such as daylight and artificial, the two should obviously be of as similar *composition* as possible. Not only will individual colours seem much the same under both illuminations, but the relationship between colours will be constant; matching pairs will continue to match, even if the dyes are metameric (page 152). When one famous chain of clothing shops refitted all its branches with fluorescent lighting which was very rich in blue, many of their goods, which matched each other acceptably in daylight and in most artificial lighting, ceased to do so in the shop; and sales duly slumped.

What of man-made objects? Most have been deliberately coloured; but to what end? Their colour may inform the intellect, or modify the emotions; and, often, it serves both ends. The rest of the chapter deals with those more prosaic uses of colour which convey information, whether by naturalistic or symbolic means.

The techniques of colour reproduction (Chapter 16) are of immense archival value. A skilfully executed, naturalistic colour photograph, like a representational painting, can provide an invaluable and widely distributed record of the object, for study as well as for pleasure. The most faithful impressions are obtained when the objects are small and fairly flat. Butterflies, pottery fragments, miniature enamels, brooches and illuminated manuscripts are particularly suitable, since most observers will favour the same viewpoint and so see with the same perspective. And if the representation can be life-size, the object and the picture will stimulate the same area of the retina and will undergo a similar interaction of neighbouring colours. To appreciate the effect of depth and scale on aesthetic impact we need only compare a statue with its photograph, a building with the architect's model, or a large picture with a post-card reproduction.

Naturalistic photography may deviate intentionally from the perfect representation. Perspective may be flattened or exaggerated and contrast increased. In the hands of a skilful advertising agent, such effects may both arrest the viewer's attention and enhance the appeal of a product, whether by association or by suggestion. Cars can look more spacious, sultanas more luscious, when the camera lies but slightly. But such manipulation of feelings are outside the scope of this chapter.

Of all the non-representational messages conveyed by colour, the simplest is 'Look', or 'Here is a boundary'; between object and environment, or between figure and ground. The more the contrast in colour, the greater the impact. In ancient Rome, names on inscriptions were picked out in red, which we still favour for warning signs, railway signals and fire appliances. Effectiveness is even greater with the fluorescent dyes used for life-rafts and safety jackets, with reflectors on vehicles, and light-emitting sources. Impact is further enhanced by rapid change: the flashing lights of an ambulance attract more attention than the more leisurely traffic lights, and 'moving' neon-light advertisements hold our interest for longer than still ones.

Colour may also be used to conceal a boundary, by use of a colour similar to the background. Alpine troops wear white; naval vessels are grey, the apparent colour of distant sea. Boundaries may be further disguised by breaking up familiar outlines with splodges of different, but drab, colours. During the Second World War, the British government sensibly sought advice on camouflage from a specialist on protective coloration in animals (cf. page 99).

The use of colour to emphasize a boundary is often extended to its use for classification. Members of particular groups, be they pupils of a school, members or supporters of a football team or tins of soup made by one manufacturer, can be recognized more easily, if they are adorned with a particular colour, or a simple combination of colours. The groups may be subdivided into smaller classes, each labelled by colour: pipes for different services in industrial installations, academic hoods for degrees in different subjects, the values on postage stamps, cylinders of gases, grammatical parts of speech in children's exercise books, lines and areas on diverse types of maps and charts, and the different categories of Penguin Book (A bright green spine would be scarcely appropriate for this one.) Colours may be used to differentiate between objects which would otherwise be identical, such as croquet balls, pills of different strengths, electrical wires, roadside reflectors.

Pattern may be invoked as well as colour, as in ships' signalling flags and heraldry, whether medieval or modern. In some situations, colours have meanings fixed by convention: red implies the port side and green the starboard, so that not only can an observer locate a boat at night, but he also knows the direction in which it is travelling. Heraldic devices are often combined to represent a complex genealogy; and since colours play so important a part, a system has been devised for recording heraldic 'tinctures' on such surfaces as engraved metal by different types of hatching (see Figure 82, page 225). Colour can contribute to, or even form the sole basis of, a fairly elaborate code. When ships communicated with signalling flags, there was a separate flag for each letter and each digit; and some single flags, or pairs of flags, were used to convey such messages as 'preparing to sail', 'in need of assistance' or 'crew mutinied'. The coloured bands on an electrical resistor can be decoded by the initiated to give both the resistance and the manufacturer's specification of its precision (see Figure 79).

We may speculate on why a particular colour, or set of colours, has been chosen for a certain job. Some are clearly representational in origin. Mountains, as well as maps, change colour from green, through grey or brown to

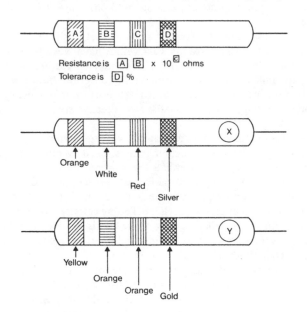

Resistance is $\boxed{A}\ \boxed{B}$ x $10^{\boxed{C}}$ ohms
Tolerance is $\boxed{D}$ %

Figure 79. Code for resistors *The digits A, B, C are represented by colours:*

0	Black	5	Green
1	Brown	6	Blue
2	Red	7	Violet
3	Orange	8	Grey
4	Yellow	9	White

The tolerances, D, are:

10	Silver	5	Gold

Decoding gives the resistance of X as 3,900 ohms ±10% and of Y as 4,300 ohms ±5%.

white as altitude increases; and (if calm) the sea is lighter where it is shallow. Since arterial blood is more scarlet (or less purple) than venous blood, diagrams of the circulation system often show those vessels which carry blood away from the heart in red, while those which return the deoxygenated blood are shown in blue. Some choices are technological. The colours used on 'instant temperature maps' are determined by the response of the dyes in the colour film used to record the infra-red radiation emanating from the subject. Some are purely practical. Green and red are an obvious choice for lights on ships, aircraft and traffic signals, since they can be seen and recognized from a considerable distance. Yellow can be seen even more easily and is much used for rescue vehicles and road markings; but the normally sighted can distinguish between green and red more reliably than between green and yellow or between yellow and red. Blue would be a poor choice for this purpose as blue light is difficult to identify from afar, although it is perfectly adequate for distinguishing emergency vehicles at short distances. Tolerance limits for coloured signalling lights are shown in Figure 80.

Colours of road signs are chosen to provide maximal contrast, both between symbol and ground and between the whole sign and the environment. Some coding colours are chosen for ease of memory, as for electrical resistance; others, as in commercial display, are purely arbitrary. Since red is widely recognized as a danger signal (derived, perhaps, from a primitive reaction to fire, or to blood), red would seem a good choice for the colour of 'stop' signals and of cylinders of so explosive a gas as hydrogen; but such subjective speculations touch on the emotional implications of colour, which are the concern of the next chapter. Some coding colours are chosen with a view to the needs of those with anomalous colour vision.

How do those with anomalous colour vision cope with the diversity of messages transmitted by means of colour? Luckily, many of these are not relayed solely by colour. Road signs are designed with high contrast in lightness, and gas cylinders are clearly labelled. The colour code for domestic electrical wiring was changed in order to help those for whom red resembles green. The old system (red for live, black for neutral and green for earth) seems a natural choice to those with normal vision, but could be lethal if green is mistaken for red. The present code (brown for live, blue for neutral and spiralling green and yellow for earth) is far less liable to misinterpretation.

Perhaps the best aid for those with anomalous colour vision is an early diagnosis. They may then be made aware of situations which are likely to cause confusion and taught to rely whenever possible on factors other than ambiguous colours: to stop, for example, when the top traffic light shines. Red-green ambiguity can be a problem for those using some coloured

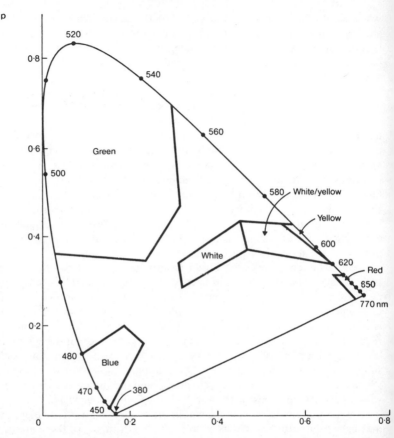

Figure 80. Coloured signals *Tolerance limits for coloured signalling lights. An acceptable coloured filter should give a colour within the approved area for a range of sources of light. (Reproduced, with permission, from D. B. Judd and G. Wysecki,* Color in Business, Science and Industry, *3rd edition, Wiley, New York, 1975.)*

indicators for chemical analysis, but, if the handicap has been diagnosed, it is often possible to choose a different indicator, which changes colour in a range to which the observer is more sensitive.

There are, however, a few occupations which are effectively barred to those who cannot tell a port light from a starboard one, or decode the coloured bands of an electrical resistor. An early diagnosis helps here, too, so that hopes are not pinned on a career at sea, in electronic engineering or, of course, in any job which involves colour matching or mixing.

19

COMMUNICATING FEELINGS

Since colour is a sensation, it would be odd if mankind had restricted its use to the communication merely of information. The stimulation of sense organs often results in emotional changes, and there is a wide variety of situations in which feelings are communicated through colour. The graphic arts are an obvious example, and also serve to pinpoint the difficulty of disentangling the effect of colour alone from the effect of other features, such as shape or tactual content. An area of blue may promote rather different moods if used in different contexts, whether, perhaps, for a sky or a human face, or even whether the area is a circle or a triangle. We meet similar problems when we try to disentangle the factors which contribute to feelings produced by the different colours used to adorn our buildings or our persons. Even the effect of coloured lighting depends to some extent on exactly what is being lit.

In any real situation, it is effectively impossible to study the impact of colour on its own. Experiments can, of course, be performed under labora-tory conditions, where a white screen, or a bare white room, is illuminated by light of a particular colour. Although feelings of any subject under such artificial conditions will probably be very different from those of the same person outside the laboratory, such work at least enables the effect of different colours to be compared. And results obtained both with infants and with the mentally ill show that red light is exciting, promoting increased bodily activity, feelings of love, anxiety or anger. Blue light, on the other hand, has a calming, or a depressive effect, compared to white light, or to light of intermediate wavelengths. These changes of mood with the colour of the illuminating light form the basis of several forms of colour therapy. Some, but not all, of these treatments contain so important an element of suggestion as to verge on the occult.

In most studies of the colour preferences of schoolchildren and adults, blue was the favourite, with red second, irrespective of sex, race or age. But babies and toddlers, perhaps as yet unable to appreciate calm, seem to prefer red to blue.

The increase in excitement on passing from blue to red has long been appreciated by those involved in interior lighting, decorating and furnishing. Blue is reckoned to be a 'cold' colour,* while orange and pink are 'warm' ones. The eye is very sensitive to the effect of small changes in the composition of 'white' light. Lighting by fluorescent tubing is said to be 'harsher' light than that from tungsten filament bulbs, which emit light with a higher ratio of red to blue. The exact colour of the light from a single tungsten bulb is also changed when a dimmer is used; the less the electricity which passes through the filament, the lower is its temperature and the redder the light. A 'dimmed' light therefore looks a 'warmer' colour than an undimmed one, as well as being less bright. A candle gives a still 'softer' light, it being both redder and even less intense.

Stage lighting in the theatre is a complex and highly technical craft in which both the colour and the intensity of the light serve to influence mood, while differences in brightness serve to direct the attention of the audience to selected parts of the stage. The composition of the light from a particular source is varied by using 'white' sources behind filters of red, blue and green. The lamps behind each type of filter are connected in three different circuits, each controlled by a separate dimmer. A modern stage set has battens of such lights in many different positions and the consoles by which they are controlled may have about 1,000 separate dimmers, which nowadays are often operated electronically. Since perceived colour is determined largely by the composition of the light entering the eye, it depends both on the composition of the light from the illuminating source and on the nature of the surface from which it is thrown into our field of vision.

Inside a building, the illumination is usually predominantly 'white', and the main colours we see are those of the interior decoration, and of the furnishings. Colour is used in a number of ways: on its own, as a substantial area of self-coloured wall, ceiling, carpet or curtain, or a smaller area of sofa, chair or cushion; as part of a purely decorative pattern, as on wallpaper or soft furnishings; or as part of a picture, in which colour may play one or more of a number of aesthetic roles (see page 213). Large areas of colour can cause appreciable changes in the light falling on other areas, especially in narrow rooms and corridors, where successive reflection of light between opposite walls makes colour seem darker. The reflection level within a room can be kept at a restfully steady level by decreasing the reflectivity of the better-lit walls.

The relationship between the coloured objects in a room may itself constitute a pattern and so contribute to the impact of the room on the observer. Not surprisingly, the dominant colour inside a room seems to

* The 'psychological warmth' of a colour has no connection with the technical definition of colour temperature. Indeed, blue has the highest colour temperature of visible radiation, and red the lowest (see page 165).

affect mood in much the same way as do coloured lights. Popular opinion deems a pink room to be 'warmer' than a blue one: and would describe yellow as 'cheerful', and green as 'restful'. Theatres often had 'green rooms' in which actors waited to go on stage. Dark or intense colours are often felt to be oppressive. Authorities of a mental hospital found that anxious patients were more likely to venture to the end of the corridor if the corridor were painted dark purple, brown or crimson than if it were painted in some lighter colour; the oppressive effect provided motivation for them to reach for some more peaceful destination. But 'peacefulness' in colour is finally a personal choice. Disturbed children in one residential home were found to settle down more relaxedly at night if surrounded by a colour which appealed to them individually. Each child was first asked to paint a picture and the small screens which made up his own cubicle were then painted in the colour he predominantly used.

The colour may be determined by tradition. In Mykonos, houses are white with all doors and window frames (like all the boats) painted turquoise. Only the church domes depart from this scheme. A wooden Swedish house is conventionally rust-red, with white window frames and a pale-green door. Present-day French architects use colour more individually. A number of different but related colours may be used to decorate, and break up, large surfaces; one wall of a block of flats, a passage between two Métro platforms, a wide pavement, the tiled surface of a pool, or the service ducts running outside the Centre Georges Pompidou. Variation in colour may also be provided by striped curtains behind plain glass, or by reflections of the clouds in the tinted mirrors which clad some of the Paris tower blocks. Temporary colour is even provided by gaily painted fencing around building sites.

We do much to alter the colour of our more 'natural' environment. Gardens range in size from a window box to a landscaped park; and, for either, our plants are often chosen for their colour: of their leaves in various seasons and of their fruit, as well as of their flowers

Colour is also of great importance in personal adornment, be it sartorial or cosmetic, but here the relationship with mood is much more complicated than the effect of a patch of coloured light or a painted wall. Clothing, in particular, fulfils so many functions that it is difficult to isolate the effect of the colour itself from its social or symbolic context. A uniform, almost regardless of its colour, may induce feelings of pride, camaraderie, envy, hostility, fear, compassion, gratitude, amusement ... The list is almost endless, and the effect depends largely on who is looking at whom: hussar, schoolgirl, prisoner, nurse, football fan, policeman, air hostess, Hitler Youth, nun. Though the actual colour has but small influence compared with the social connotations of the uniform, the author surely cannot be alone in feeling a pang of sympathy for officers of those regiments whose mess-tunic is other than red. That such feelings are as likely to stem from an early exposure to toy soldiers as from any physiological effect of scarlet merely illustrates the difficulty of trying to disentangle any inherent psychological property of colour from our reaction to uniforms.

Colour in clothing may have strong association with particular occasions. In the Roman Catholic Church and the Church of England, priests wear vestments of green, red, purple, black or white, depending on the Christian calendar, and use matching furnishings for the altar and lectern. The colours of academic dress vary with the university, the faculty and the seniority of the degree. In English universities, the everyday gown is black, derived from the black habit of the Benedictine friars, but St Andrew's University in Scotland wears scarlet. Academic hoods, and the robes used for university festivals, come in a bewildering variety of combinations of colours, with or without fur trimmings. The use of colour to denote earned distinction is widespread. Tyrian purple was only for the great: the High Priest in the Temple of the Israelites, for the robes of Darius and of Alexander the Great, for the sails of Cleopatra's barge. Caesar and Augustus decreed that none but the Emperor might wear it, and in Nero's day even the selling of it was punishable by death. Hand-tattooing of Maori tribesmen was performed only on those of proven ability as warriors. Our own society rewards military exploits with medal ribbons, equestrian excellence with rosettes and sporting prowess with 'colours'.

Dress of a particular colour is sometimes worn for reasons which are more associative than heraldic. Our present culture considers black, or sombre colours, appropriate for mourning and white for such occasions as weddings, first communion and confirmations. In Europe, the use of yellow for women's clothing has often carried suggestively sexual overtones, although it is the colour adopted by monks in the East. And English schoolboys often avoid wearing pink or mauve lest their contemporaries think them to be homosexual. Even when account has been taken of the many conventions and associations of colours in clothing, considerable scope for personal factors remains. So much of the impression created by clothing and make-up depends on both the wearer and the observer. A woman in black may look alluring, depressing or business-like. The effect of red clothing, like red cheeks or red lips, may be pleasingly cheerful, or displeasingly flashy. But, stripped so far as possible of any conventional or associative significance, it is probably true that for dresses, as for lights, blue and green look more restful than red or orange.

By far the most complex use of colour to communicate feeling is in the graphic arts. This book is no place in which to attempt to analyse aesthetic experience, nor yet to explore the various parts played by colour throughout the long history of art. Colour has, of course, been used to create images, patterns and abstract compositions in a large number of contexts. It is generally assumed that the hunting scenes of primitive representational art had their origins in the magic or religious practices of the day and were maybe used in rites designed to further success in hunting and hence to ensure the very survival of the community. Although it is less easy to account for embellishment by non-representational pattern, the urge to decorate runs very deep. Not long ago, an old Cretan villager used to sit under an olive tree at Agia Triada making bamboo 'pipes of Pan' to sell to the tourists. When he finished one, he submitted it to rigorous testing. If it was musically acceptable, he sliced off the shiny, corrugated front and decorated the new flat surface with wavy blue-black lines from a plastic ball-point pen. He seemed in no doubt that the pipes were thereby much enhanced; and most people, at most times in history, would probably have agreed with him. Our own taste, conditioned to enjoy the weathered stonework of a Greek statue or northern cathedral, must surely represent the minority view, and would doubtless have amazed those who painstakingly applied the original decorations in all their polychromatic splendour.

Our forebears might also have been surprised by our food technologists' uses of colour. The pastel green and blue piping on a white iced cake might be so unpalatably suggestive of mould as to destroy any pleasure in the decoration.

Naturally, it is difficult to draw a clear line between pattern and formalized representation; although some of the intertwined Islamic tile designs might seem purely decorative, the predominance of green and blue have led to the suggestion that they represent streams amid fertile land. Since the need of desert peoples for water is as great as that of the northern cave dwellers for success in hunting, these traditional Islamic motifs may also have as their origin some visual plea for survival. This is, of course, speculation; but we do know for certain that man has long decorated both his dwellings and his artefacts, and that, whatever their origins may have been, visual arts have in their time fulfilled many different, and complex, functions. The rich decoration of many places of worship is a testimony to man's need both to honour his gods and to receive proper recognition for the financial success and pious generosity which makes such splendour possible. The explicitly representational element in much Christian art arises from its use as a teaching aid in times of mass illiteracy. Would that the educational aids of today had the aesthetic impact of the windows of Chartres: over 2,000 square metres of gem colours of an unimaginable brilliance.

In domestic and civic contexts, painting has been widely used for portraiture, mainly of people, although animals, flowers, fruit and even buildings have been lovingly recorded, thereby defying immediate oblivion. As a legacy from our cave period, such paintings may still contain a deep-seated plea for some form of survival. Many post-Renaissance narrative paintings were concerned with non-religious narratives, often from classical mythology. Later still, historical, domestic, naval and military events were depicted, with varying proportions of imagination and realism. Towards our own time, painters, joined latterly by photographers, were concerned to exploit the full power of their art to convey the effects of light and atmosphere. The visual arts are so complex, and so diverse, in purpose, subject and technique, that they resist attempts at tidy classification, or even at general explanation. While the cynic might insist that art arises from our need for status symbols, the idealist would surely stress man's urge to enrich even the humblest aspects of his environment. Today, both artist and philistine might agree that much modern art arises from the need of the artist to express himself in a coherent way which is (moderately) acceptable to society. And not one of these views can be proved wrong. Whatever the forces which drive the artist, added colour is not an essential element in the production of visual art. But it is none the less an extremely important one.

As we have seen, colour can be applied in a wide variety of ways to an even wider variety of objects. Here we shall confine ourselves to the use of colour by artists, usually by those who work on surfaces, be they flat or curved, and who would aim to produce 'art' rather than 'decoration', although they might find it difficult to draw a satisfactory line between the two. For the moment we shall exclude art with any component which moves, and so confine ourselves mainly to 'pictures', be they paintings, mosaics, appliqués, stained-glass windows, enamelled miniatures, collages or tapestries. There are, of course, vast differences between the effects of pictures created from different materials. But within a single technique, there is much scope for variety. Colour can be used in many different ways and for many different purposes. Some of the reasons why artists use colour emerge if we compare an illuminated manuscript with an early printed book, a charcoal drawing with a work in pastel, an engraving with an aquatint, a sepia photograph with a colour print, a pebble mosaic in black and white with one over the whole natural colour range. Even in these few examples, we may see colour being used in attempts to replicate nature, to emphasize the difference between one area and another, to produce contrast, to give a sense of depth, to add brilliance, to provide information by symbolism, to generate a mood. Not that all the coloured versions are always preferred by all viewers. But colour indisputably provides additional aesthetic scope, both for success and disaster.

In attempts to describe different types of colour experience, some psychologists have drawn up such classes as 'film' colours (structureless patches of uniformly coloured light on a plain screen), 'volume colours' (as in a glass or flask of transparent coloured liquid) and 'surface colours' (caused by light scattered from an opaque surface, whether flat or curved). In this scheme, craftsmen in such transparent materials as precious stones, glass and plastics can exploit the volume colour of solids. Some present-day artists in stained glass incorporate in their work blocks, rods and rough chunks of glass, of very different thickness. Others use partially overlapping layers of glass of different colours to give an illusion of solidity.

Most artists would be classed as using surface colour in that the colouring matter itself is either opaque, or, if transparent, is spread in a thin, even layer over an opaque surface. But the texture of the surface may markedly modify its appearance. A coat of varnish produces highlights of white light or of very unsaturated colours. Light may be reflected in different directions from angled tesserae of wall mosaics, from the sparkling flecks in metallic 'inks', from smooth gold leaf, from the different brushwork textures or palette knife with which oil paint may be applied.

One of the many difficulties about commenting on the artist's use of colour is its relatively transitory nature. Little remains of the paint which adorned classical and medieval stonework, and it is difficult to imagine the original colours of tapestries which now seem to have been made from thread dyed to different depths of either brown or indigo (reds and yellows having faded to browns, and green to dull indigo). Varnishes often yellow or brown with age, and surfaces become begrimed (cf. page 189). Judicious removal of a superficial layer of dirty varnish can transform the murkiness of many an Old Master to a brilliance which may seem almost garish to the unaccustomed eye.

Although many paintings suggest that artists throughout the ages delighted in colours as bright as they had available, this was not universally so. Indeed, colour has been put to a wide variety of uses, realistic and otherwise. The colours might have symbolic significance, sometimes almost as explicit as a colour code. Minoan painters represent girls as white-skinned but boys as chestnut; and during the Italian Renaissance the Virgin Mary was usually robed in blue. Perhaps colour's most basic use is to differentiate one area from another, as in a political map, and more decoratively, in Greek vase painting, stained-glass windows, and much of Matisse's work. Lines of a different, often darker, colour are often used to emphasize outlines and to supply added detail. Many examples may be found in pictorial stained glass, Minoan frescoes, pebble and Byzantine mosaics, Russian icons, medieval manuscript illuminations, much folk-art, and the paintings of Georges Rouault and L. S. Lowry. Black outlines also serve to keep colours purer by preventing optical mixing (page 136) at boundaries, as do the white spaces between areas of paint in the work of Delacroix and Cézanne.

An illusion of solidity can be created by using darker colours to represent those regions which are less well illuminated, and by allowing these shadowed areas to change gradually into those of brighter colours, which themselves are shaded into highlights of very unsaturated colours, or even of white. The use of colour to create a feeling of depth is as old as painting itself; the animals of the cave-painters have considerable solidity. Later artists, of course, had more colours at their disposal, and for those who sought to paint more or less naturalistically, shaded areas posed a particular problem. What colour is a shadow? Indeed, what colour is an object? We have seen (Chapter 14) that the psychologists of colour vision have detected two opposed tendencies in observers. One is an awareness of contrast. As we scan our field of vision, our response to one region is modified by our response to neighbouring areas in a way which (usually) emphasizes any difference between them. On the other hand, when we see familiar objects, our interpretation of any response is modified by what we should see. We expect a real donkey always to look donkey-coloured (see page 140) regardless of how it is lit. But we are more open-minded when looking at a painting. We know that paint can be of any colour and that present-day artists are not past making animals any colour; what about Marc's *Tower of Blue Horses*?

Until fairly recently, many naturalistic artists have painted objects mainly in 'local colour', a technical term implying the colour they would appear in white light. Shaded areas were usually depicted by darker, predominantly neutral colours, such as browns and greys. By grading colours in both saturation and brightness, remarkable effects of solidity can be achieved. Nor was the illusion of the third dimension restricted to solids. Many Renaissance and later painters used graduation of hue in landscapes to give an impression of distance, often achieving their 'aerial perspective' by using brownish hues in the foreground, greens and yellows in the middle distance, and blues and greys in the far distance. Even a totally non-representational painting can have a feeling of depth engendered by colour alone, without any reinforcement from linear perspective. Reds, yellows and browns give the impression that they are advancing towards the viewer, while greens and blues recede into the picture (see pages 66 and 130).

The texture of a painting plays its part, too. Artists who use oils may apply one or more layers of transparent paint, over an opaque ground; or the ground may be covered by a second layer of paint which is semi-opaque, or opaque but applied so as to give only incomplete coverage. Whatever the technique, some of the ground shows through the layer of transparent 'glaze', or opaque 'scumble', and the colours of the different layers modify each other. Layers of water colour may be similarly applied, as 'washes'. In oil paintings particularly, the surface layer seems to approach and the ground to recede. The rich colour in many of Rubens's works owes much to his use of series of glazes, and Turner achieved his atmospheric effects partly by applying light glazes over light grounds.

However, not all painters exploited colour to this extent. Some positively sought to work in a subdued range of hues in order to get the fullest possible impact from variations in intensity. Colour plays only a small part in Rembrandt's dramatic use of chiaroscuro. Some painters, such as Claude Lorraine, even viewed scenes through a darkened lens so that they could appreciate variations of intensity without being distracted by differences in hue, a problem which present-day photographers in black and white try to solve by self-discipline. The restrained colours associated with the 'Old Masters' cannot be wholly ascribed to surface grime. And artists may shun bright colours through unconscious preference as well as by conscious acts of restraint. When Goya became deaf, and depressed, at the age of forty-seven, he started to paint in unusually sombre colours.

The tradition of local or less-than-local colour and subdued shadows gradually gave way to the practice of trying to replicate the composition of light entering the eye. That part of a white vase close to a yellow wall will, itself, look yellow. A white horse, cut out of the turf in a chalk down, will, by simultaneous contrast with the grass, appear pinkish. And, as Delacroix emphasized in both words and paint, sand in shadow by contrast with the same sand in sunlight will look purple. Shadows on pink similarly look green. In Frank Kupka's picture of a young girl, sunlit flesh was painted yellow, shading to purple, while those parts in shadow were represented as green, shading to pink. In Robert Delauney's self-portrait, much of the face is painted a deep sea-green, with prominent purple cheek-bones. Both paintings seem perfectly plausible. Anyone sceptical that healthy flesh can be convincingly painted a brightish green need only look at a Renoir, preferably a real one, because a reproduction will usually be scaled-down in size. Obviously, as colours are modified by adjacent colours, any change which alters the number of frontiers between coloured areas in our field of vision will alter the effect of the colour stimulus we receive. So no reproduction, however well the individual colours are copied, can ever be a perfect replica unless it is the same size as the original.

The way in which one colour is influenced by its neighbours also depends, of course, on the size of the actual areas of pigment which an artist uses. We have seen that, if the patches of paint are small enough, they cannot be resolved by the eye, and that the whole visual colour range can be reproduced by printing small dots in inks of not more than four colours (page

184). Light of different colours from the four pigments is perceived by the eye as light of a single 'colour'. Impressionist painters used this same principle of 'optical mixing' to control the composition of the light which enters the eye. Very small, closely spaced dots of green and red would, when viewed from an appropriate distance, give an effect of yellow: and in skilled hands, a more vibrant yellow might be achieved than by traditional methods. The Impressionists, unlike many printers, did not restrict themselves to four colours; nor to the circular dots favoured by Seurat and Signac. Monet used dabs and short dashes, and Toulouse-Lautrec longer lines. Van Gogh's slashes, which were still longer, often sinusoidal, could, at their best, suggest movement as well as light. The Impressionists often worked with bright colours, which varied widely in hue but little in brightness. Their pictures, though full of light and atmosphere, can seem unsubstantial. They lack the difference in intensity of reflected light between sunlight and shadow. It needed Cézanne's reintroduction of darker colours for shaded regions to restore the illusion of depth.

In our present century, some artists think paintings need no longer be even remotely naturalistic. They feel free to experiment with different shapes and juxtapositions of coloured areas and explore the territory of the psychophysicist: the subtle relationship between simultaneous contrast, colour constancy, recognizable shape, intricacy of design and total area. The work of Sydney Harry emphasizes the changes which occur with viewing distance. As optical mixing takes place, the hues change; and so, too, does the brightness, and hence the relative importance of the design elements. His patterns change shape as you approach them. Much of the work in the early part of the century was linked to theoretical studies of colour, by both artists and scientists. It resulted in the evolution of the many colour maps and solids discussed in Chapter 15. Kandinsky tried to establish a connection between such colours as red, blue or yellow and simple geometrical forms, and also observed that yellow, as well as 'advancing' towards the viewer, also appears to spread out, while 'receding' colours, like blue, contract (see page 131). But attempts made on any quantitative basis to lay down rules for, or even to draw up a code of, colour harmony have as yet been unsatisfactory. Those colours which 'go' with other colours, whether or not in the same range of hue, are largely a matter of personal taste, which is doubtless compounded of individual preferences and current fashion. Some decades ago, the juxtaposition of orange and magenta would have seemed discordant to many people, and worn only by the severely eccentric. But the vibrant simultaneous contrast was then exhibited in much 'Pop' art, with the result that our visual system has become acclimatized to finding orange and magenta a familiar, and almost acceptable, combination. The very use of such 'aural' words as 'harmony', 'clash' and 'discordant' in discussions of colour relationships suggests an assumption that our senses find some combinations of colours more pleasing than others and that such preferences can be generalized for colour, as for sound, according to mathematical criteria. But, as we shall see, the relationship between music and colour is somewhat tenuous.

20

COLOUR, MUSIC AND MOVEMENT

It seems that infants do not keep their senses in sealed compartments. A loud noise, for example, may produce a sensation not only of sound, but also of colour. Some of us may even remember having such synaesthetic experiences, often connected with particular words (see page 222), or with music. As the child gets older, the distinction between sensations of sound and of colour usually becomes absolute; but some adult musicians continue to associate a colour with a particular note, timbre or musical key. Berlioz, Debussy and Wagner were particularly sensitive to sensations of colour in music. Most of those who have experiences of this type agree that sharp keys are 'brighter' than flat ones, but association with a particular hue seems purely personal. Thus the key of C major was red to Scriabin, but white to Rimsky-Korsakov.

Throughout the centuries, many attempts have been made to find a mathematical relationship between music and colour; one of the earliest was that of Ptolemy in the second century A D. It is not surprising that no correlation has been found. The sensation of sound is caused by the action of oscillations of air pressure on the ear, while that of colour is due to the impact on the retina of electro-magnetic oscillations, of a frequency more than 10,000 million times the frequency of soundwaves. There seems to be no recorded instance of the converse phenomenon: colours do not seem to produce sensations of sound.

Even if most of us no longer experience synaesthetic sensations, the association between colour and music is often very pleasurable. Bliss wrote a *Colour Symphony* in which he described the mood of each movement by a colour. Light of the appropriate colour was to be projected on to a screen during performance; but the audience felt that the music gained little from the addition of the colour. Scriabin's *Prometheus – The Poem of Fire* was also performed to a play of coloured lights, but the chromatic effects were ill-received by the critics. Several ingenious mechanical devices have been used to generate colour changes synchronized with music. The 'colour organ', for example, produced a different colour on the screen for each note struck. Coloured areas of varying shape and intensity, as well as of different hue and duration, could be produced by cartoon cinematography, as in the abstract movements of the Walt Disney film *Fantasia*. Mary Ellen Bute used, in addition, electronic and optical methods to create abstract colour films relating visual and aural sensations. Similar displays of three-dimensional holographic images have since been presented in 'laserium concerts'. Although these examples may seem to be primarily of collectors' interest, much of the impact of ballet, too, depends on the combined sensations of music and of moving colour, as does the use of 'psychedelic' lighting in a modern dance hall or disco.

Descendants of the colour organ are devices such as the Clavilux for producing colour patterns which change rhythmically in time *as if* to music. More distant relatives of such instruments are objects of kinetic art and some commercial and domestic lamps where the colours change at a constant rate, and so any rhythmical connection with music has been severed. But there remains the fascination of change, the merging of one colour into another. Whether we are watching the rising sun or admiring a piece of kinetic art, the element of change can do much to enhance the sensation

21

WORDS AND COLOURS

Synaesthetic colours are not limited to music:

> When we were children words were coloured
> (Harlot and murder were dark purple) ...

Many of us doubtless remember similar sensations of colour, perhaps, like Louis MacNeice, associated with individual words, or perhaps as a way of distinguishing between members of a small set of words, or the numbers one to nine. The origins of such associations may lie in some object encountered at a very early age, such as a book or board game. For the author's sole example of coloured words, any origin in the real world seems unlikely; who would even have printed a calendar using so many shades of one colour as:

Mon.	*Tues.*	*Wed.*	*Thurs.*	*Fri.*	*Sat.*	*Sun.*
Pale blue	Cherry	Gold	Dark chocolate	Chestnut	Ochre	Black

The memory is vivid still, confirming the luminosity of Wednesday compared with the apparently similar, but much drabber, Saturday. More interesting, though, than such idiosyncratic associations is the role played by words denoting colours.

Studies of colour terms in a variety of languages from unwritten, primitive dialects to modern European languages suggest that the development of a vocabulary of colour follows an almost regular pattern. First, colours are classified as either 'white' (or pale) or 'black' (or dark). The first hue to be distinguished is invariably red. This is followed by yellow and green, although not necessarily in that order. When both yellow and green are established in the language, blue is added, and then brown. The remaining four terms, which complete the basic eleven-term colour vocabulary of modern European languages, may have been added in any order, giving Berlin and Kay's evolutionary scheme (Figure 81), which seems to be quite general. So, if it were found that a particular primitive people divided coloured samples into only five named categories, these would roughly correspond with black, white, red, yellow and green. Blues would be distributed among samples classified as 'white', 'green' and 'black', depending on whether they were pale, bright or dark blues. Even in languages which have words for both blue and green, individuals find it particularly difficult to distinguish between them. The English seem to indulge in frequent discussions as to whether various turquoise objects are 'really' blue or green.

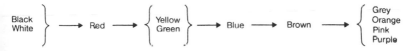

Figure 81. *Evolution of colour words. Berlin and Kay's scheme*

Studies on the number of colour words used by young children in Russia suggest that the terms are added to the child's vocabulary in the same order as they were added in the course of evolution of adult language. It would be interesting to know if this were also true for children in other cultures.

There is, of course, no justification for the assumption that, just because a people have no word for a colour, they are unable to see it. Gladstone asserted that the ancient Greeks were insensitive to blue, Homer uses no such word; and there has been much scholarly speculation about exactly what Homer meant by his 'wine-dark' sea. Homeric Greek does indeed have few words describing hues. The term for green has alone survived for current everyday use; and that can serve to describe alike the vegetation and the sky-blue national buses. The modern Greek colour words for blue onwards are foreign loans or have been taken directly from nouns such as 'orange'. Many languages lack any generic term for brown; in modern Greek, suntan is 'black', all brown hair is 'chestnut', and other appropriate browns are 'coffee'. Some colour words have undergone dramatic changes in meaning; others have long been imprecise. The Homeric *kuan*, which implied dull or pebble coloured, has undergone a marked change of meaning to become the vivid printers' primary colour 'cyan' (see page 179). Classical 'purple' implied dyed with murex, whether the cloth had a newly dyed brilliance or had faded to a dull brown. In old High German, 'brown' meant either 'gleaming' or 'dark'; to Dr Johnson 'brown' was any colour mixed with black; and today it is 'a highly unsaturated yellow' (see page 165). 'Coffee' as an English colour word surely varies widely with the amount of milk preferred by the individual; and the Welsh *glas* refers to the colour of mountain lakes, whether blue or green.

Nor does a specific colour name make that colour any the more visible. How many children must have searched the rainbow in vain for the indigo they were told they could see? Authorities have varied in the number of distinct colours of the rainbow. Aristotle settled for red, green and blue, sometimes with yellow. But Seneca, and Newton in his early work, favoured an infinite gradation of colours. In his *Opticks*, however, Newton chose to recognize seven separate colours, in order to draw an (inappropriate) analogy with the musical notes A to G of the octave scale. Although he named these as red, orange, yellow, green, blue, indigo, violet, he probably could not identify them unequivocally himself. The spectrum he obtained from sunlight was described by an 'assistant, whose eyes for distinguishing colours were more critical than mine'. It seems likely that Newton's blue corresponds more to our cyan, while his indigo is a deeper, purer blue. An eighteenth-century poet, Richard Savage, writes in his poem 'The Wanderer':

> All-chearing green, that gives the spring its dye;
> The bright transparent blue, that robes the sky;
> And indigo, which shaded night displays,
> And violet, which in the view decays.

But indigo is tenacious in its dubious claim to rainbow component status. A British postage stamp of 1981 shows a completed painting of a formalized primary bow: of the seven Newtonian colours in bands of exactly equal width.

Studies on colour words in literature are not restricted to ancient authors. Havelock Ellis surveyed twenty-five literary sources, ranging from the Bible and the Mountain Chant of the Navajo Indians to Shakespeare as well as many nineteenth-century English and French writers. He noted the frequency with which each of seventeen colour terms was used. Later workers, aided by computers, have continued this type of work. Although it is interesting that such a nature-lover as Wordsworth mentioned green twice as frequently as any other colour, it is more difficult to relate the findings to other aspects of a writer's work. To what extent is Edgar Allan Poe's marked preference for yellow really indicative, as Ellis claims, of his turbulent mind? And can any conclusion be drawn from the fact that Blake seldom mentioned red?

We recognize, of course, that in any language colour terms often have associative meanings which may have little to do with visual perception. Some associations are not difficult to track down. 'Green' can suggest mild lushness, or an unhealthily pallid complexion, or unripeness, or something not yet desiccated. If someone can look sick with envy, he can be so jealous as to be 'green' (or so scared as to be a cowardly 'yellow'). Unripe green can become 'untried', and undesiccated imply 'vigorous'. When Horace described patches of fine writing as 'purple', he associated ornate literary style with luxury dress.

Some of the associations of colour words are compatible with our emotional responses to lights of different colours (see page 209). Red stands for desire, anger and revolution; and blue for constancy, both in love and in politics. The Greek national colours, blue and white, symbolize peace and freedom, and modern Greek poetry abounds with images of sky and cloud, of sea and foam, of blue butterflies on white flowers, of blue shutters on white houses, and of blue eyes. But blue has numerous other associations. The complexion can become blue with fear, cold, despondency. But 'blue' can also imply obscenity, or drunkenness. The German *blau* can also mean drunk, and it is claimed that alcohol causes visual images to become bluer and more distant than normal. The oath 'By all that's blue' derives from the French *parbleu*, a euphemism for *par Dieu*. If expressions involving colour terms arise from specific incidents, it is less easy to infer their meaning. The American 'yellow' press took its name from the publication, in 1895, of a picture of a girl in a yellow dress on the cover of the sensational newspaper, the *New York World*. The term *conte bleu* for a fairy tale arises from a collection of medieval adventure stories, published as the *Bibliothèque Bleu* on account of the blue covers.

In some spheres, such as heraldry, the colour terms are restricted to that particular context. The words for heraldic 'tinctures' derived from the French are shown in Figure 82.

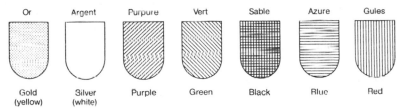

Figure 82. Heraldic tinctures *Names of colours used in heraldry, together with conventional hatching used for their representation in black and white, or on engraved objects.*

Other specialist activities use the normal colour terms, but in a way which counteracts accepted usage. A huntsman in a red coat on a white horse is described as wearing hunting pink and riding a grey. A white Siamese cat, with grey eyes, nose, feet and tail, is described as 'blue-pointed'.

Another type of difficulty arises when a wide range of colours is available, e.g. for architectural or textile design. Names for colours should certainly be informative; but attempts to make terms appealing often detract from their usefulness. The words are often mere *aides-mémoire* on a sample card. Customers will choose by colour rather than by such names as 'Waltz' or 'Charade'. Without Figure 83, can we guess even the hue?

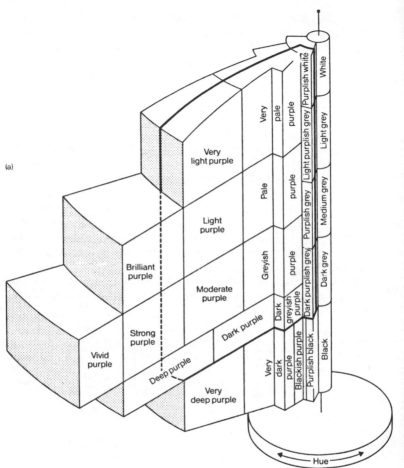

(a)

African Violets

Thistledown	Chiffon	Valentine	Capella
Colette	Wistful	Rhapsody	Nimbus
Wistaria	Veronica	Ballade	Turtledove
Waltz	Galliard	Empress	Highland
Erica	Fuchsia	Ophelia	Mephisto
Pansy	Sovereign	Wineberry	Charade

(b)

Mail orders, before the ubiquitous colour illustrations, presented greater difficulty. 'Almond' probably means green, but of the leaves or the husks? What about the pink petals, the tan outer skin of the nut, or the creamy kernel? 'Aegean' must surely be some kind of blue (or 'wine'?), depending on the depth of the water, though the Aegean can, like any other sea, look unappealingly grey.

Verbal distinctions between various shades of a single colour should be as unambiguous as possible. The nouns best suited to be colour terms are those which refer to objects which are a single and constant colour, and well known throughout that part of the world in which the language is used. Thus, sulphur, lemon, straw, canary, gold and egg yolk would seem satisfactorily exact descriptions of various types of yellow. It is less easy to envisage the precise colours of cream, ochre or mustard; each of those may have one of a range of colours. Names of flowers are best avoided in a language, such as English, which has a geographical distribution so much wider than that of the botanical species itself. Would 'primrose', 'buttercup' or 'jonquil' mean the same to readers in the United States, Britain, South Africa and Australia?

More reliable systems of colour names are based on the classification of colour by hue, saturation and brightness. The different colours exposed by cutting a section through one of the colour solids mentioned in Chapter 15 may be given prosaic, self-evident names, as well as the letters and numbers which constitute the 'map reference'.

The ISCC–NBS* system divides the whole colour solid into 267 compartments, some of which are shown in Figure 83. Within each compartment, colours are further divided and named. A dictionary lists the names of 7,500 colours both in alphabetical order and by 'map-reference' designation.

So far, this chapter has been concerned with the relationship between particular words and particular colours. There is also the problem of the large number of words such as hue, saturation and brightness (see Chapter 15) which are used to describe attributes of colour. Different sets of such technical terms have been devised for use with different systems of classification of colour, such as those devised by Ostwald and by Munsell. Other definitions are needed by those concerned with colour technology, in

---

* Inter-Society Color Council – National Bureau of Standards.

Figure 83. Names for colours *(a) A section through the ISCC–NBS colour solid. (b) 'African Violets'. Names taken from a paint manufacturer's catalogue in the approximate range outlined in (a) above.*

dyeing, printing, painting and glazing, and still others by those concerned with the physics and chemistry of the interaction of light with matter. Many of the terms needed in these various contexts are defined in Judd and Wyszecki's *Color in Business, Science and Industry.*

And what of the word 'colour' itself? Its traceable ancestry is short, reaching back only to Old French. 'Hue', on the other hand, can be traced back, through Anglo-Saxon, to a Sanskrit term signifying 'complexion', 'surface', 'appearance' and even 'beauty'. Some languages, like biblical Hebrew, have no abstract word denoting 'colour', despite a relatively rich supply of adjectives for particular colours.

Despite its short etymological history, colour as an idea has had a turbulent past, and has engaged the attentions of thinkers of such variety and distinction as Aristotle, Pliny, da Vinci, Descartes, Newton, Goethe and, in our own century, Wittgenstein. The physical aspects of colour could not, of course, be fully disentangled without an understanding of the nature of light, and ideas about colour fluctuated with the fortunes of the wave and corpuscular theories of light. The relationship between colour and white light caused much controversy. Despite Newton's observation that white light can be split by a prism into colours which can then be recombined, Goethe persisted in the Aristotelian view that colours arise by the mingling of white light with darkness. There was also disagreement about the importance to be ascribed to the observer. Leonardo da Vinci adopted the common-sense approach that there are six primary colours, white, black, red, yellow, green and blue, because these are the colours which we perceive as being totally different from each other. Newton's main interest in colour was concentrated on the nature of light; he appeared to ignore the part played by the human eye and brain. Goethe, on the other hand, fully appreciated the importance of the observer in perceiving colour, though he was far from understanding the behaviour of light.

Today, we accept that colour is a sensation, and modern philosophers, such as Wittgenstein, are accordingly concerned with it as one of the problems generated by our perception of external phenomena. Even if we lack the philosophical training to appreciate such problems, we are aware that a number of different factors contributes to any sensation of colour. The sensation of colour arises when light enters the eye, and stimulates the retinal cells to produce a response which is transmitted to the brain. One cause of this sensation is the composition of the light entering the eye, and this depends both on the composition of the light generated by a source and the way in which the light has been modified by the object we are observing. Such changes reflect, of course, both the structure of the surface and its chemical composition. But the sensation we experience also depends on the way in which the retinal cells respond to the light of different wavelengths, and this depends on a number of factors in addition to the composition of the light entering the eye. Not only are sensations of colour affected by illness, injury and drugs, they depend on the area and shape of the object, its distance from the eye, its position relative to the eye, the intensity of the light, and the colours generated by the rest of the visual field. Nor is the sensation dependent only on what is in front of the eye at one particular instant. The sensations of up to twenty minutes earlier may influence visual perception. And, since our experience of colour is affected by our memory, by our knowledge of what colour some object 'really' is, we could claim that our colour sensations are, in a sense, influenced by a lifetime's visual experience.

A less extravagant, and indeed indisputable, claim is that there is much more to colour than the absorption spectrum of a coloured transparent object or the reflection spectrum of an opaque one. There is more even to the phenomenon than the combination of these qualities with the spectral composition of the light source. It is tempting to think that the objective component of colour, the interaction of energy with matter, is now fully understood; and indeed, if we can properly envisage the nature both of electromagnetic radiation and of the electron, it may not be much of a step to grasp the way in which one affects the other. It is certain that the physics and chemistry of happenings outside the human body are better understood than the processes which occur inside. If colour vision were restricted to the matching of coloured lights in a laboratory, an explanation in terms of three types of receptor with maximal sensitivity for red, green and blue light would seem perfectly satisfying. But, in real life, colour vision is much more complex. A complete hypothesis must, in order to account for such features as optical constancy and simultaneous contrast, take note of the effect of shape, size and memory on the perception of colour.

It may be some while yet before the physics and chemistry of such processes are unravelled. And still further into the future lies the task of interpreting the effect of colours on feelings. What chain of interactions between energy and matter makes us feel calmer in blue light than in red? And how far will chemistry and physics be able to help us to understand the appeal of a painting?

INDEX

Numbers in italics refer to quotations.

ACKNOWLEDGEMENTS

Permission to quote lines from the following works is gratefully acknowledged:

To Holt, Rinehart & Winston, Publishers, for 'The Middleness of the Road' and 'Fragmentary Blue' by Robert Frost, from *The Poetry of Robert Frost*, edited by Edward Connery Lathem. Copyright 1923, 1947, © 1969 by Holt, Rinehart & Winston, Publishers.

To the Literary Trustees of Walter de la Mare and the Society of Authors as their representative, for 'Silver' by Walter de la Mare.

Library of Congress Cataloging in Publication Data

Rossotti, Hazel.
 Colour.

 Includes index.
 1. Color. I. Title.
 QC495.R87 1984   535.6   84–11451
 ISBN 0–691–08369–X (alk. paper)
 ISBN 0–691–02386–7 (pbk.)